Molecular Medicine - New Insights

Edited by Xianquan Zhan

Published in London, United Kingdom

Molecular Medicine - New Insights
http://dx.doi.org/10.5772/intechopen.1005997
Edited by Xianquan Zhan

Contributors
Haiyu Shen, Hossein Azizi, Lamarana Jallow, Na Li, Nima Ghasemi, Nosipho Treasure Mdluli, Ousman Bajinka, Sumera Khan, Thomas Skutella, Wenhao Sun, Xianquan Zhan, Xianquan Zhan, Yun-Chen Chang

First published in London, United Kingdom, 2026 by IntechOpen
IntechOpen is the global imprint of INTECHOPEN LIMITED, registered in England and Wales, registration number: 11086078, 167-169 Great Portland Street, London, W1W 5PF, United Kingdom

For EU product safety concerns: IN TECH d.o.o., Prolaz Marije Krucifikse Kozulić 3, 51000 Rijeka, Croatia, info@intechopen.com or visit our website at intechopen.com.

British Library Cataloguing-in-Publication Data
A catalogue record for this book is available from the British Library

Molecular Medicine - New Insights
Edited by Xianquan Zhan
p. cm.
Print ISBN 978-1-83635-250-1
Online ISBN 978-1-83635-249-5
eBook (PDF) ISBN 978-1-83635-251-8

If disposing of this product, please recycle the paper responsibly.

Meet the editor

Xianquan Zhan received his MD and Ph.D. in Preventive Medicine at West China University of Medical Sciences. He received his postdoctoral training in oncology and cancer proteomics at Central South University, China, and the University of Tennessee Health Science Center (UTHSC), USA. He worked at UTHSC and the Cleveland Clinic from 2001 to 2012, achieving the rank of Associate Professor at UTHSC. Currently, he is a Full Professor at Central South University and Shandong First Medical University, and an advisor to MS/Ph.D. students and postdoctoral fellows. He is also a fellow of the *Royal Society of Medicine, Word Academy of Productivity Sciences*, and *European Association for Predictive Preventive Personalized Medicine* (EPMA), a national representative of EPMA, and a member of *Sigma Xu The Scientific Research Honor Society, American Society of Clinical Oncology* (ASCO), and the *American Association for the Advancement of Sciences* (AAAS). He is also editor in chief of the *International Journal of Chronic Diseases & Therapy*, an associate editor of *EPMA Journal, Frontiers in Endocrinology*, and *BMC Medical Genomics,* and a guest editor of *Mass Spectrometry Reviews, Frontiers in Endocrinology, EPMA Journal,* and *Oxidative Medicine and Cellular Longevity*. He has published more than 190 articles, 30 book chapters, 10 books, and 6 international patents in the field of clinical proteomics and biomarkers.

Contents

Preface

Molecular medicine is developed based on a series of biomolecules, including DNAs (genes), RNAs (transcripts), proteins (proteoforms), and metabolites in the human body, and these molecules can be large-scale identified with multiomics, including genomics, transcriptomics, proteomics (proteoformics), and metabolomics. Moreover, complex biomolecular modifications occur in these biomolecules, including over 10 types of DNA modifications, over 170 types of post-transcriptional RNA modifications, and 400-600 types of post-translational protein modifications; alternative splicing also occurs in RNAs. These biomolecular modifications can be identified on a large scale with the corresponding omics. The obtained multiomics data can be used to construct signaling pathway networks to provide insights into disease mechanisms, discover effective therapeutic targets/drugs, and identify molecular pattern biomarkers for predictive, preventive, and personalized medicine (3PM) and precision medicine (PM) in disease. This book focuses on the contributions of single-omics and multi-omics to molecular medicine, which will benefit human health and healthcare, and on the discovery of novel therapeutic targets and drugs, as well as effective biomarkers for disease treatment.

The book includes scholarly contributions from various authors and is edited by a group of experts in molecular medicine. Each contribution comes as a separate chapter complete in itself but directly related to the book's topics and objectives.

It includes five chapters dealing with the topics: (1) Immune-Related Gene Alternative Splicing-Based Biomarker-Sets for Lung Adenocarcinoma; (2) Exploring the Multifaceted Role of Wnt Signaling: From Stemness, Development, and Disease to Therapeutics; (3) Cardiovascular Disease Genetic Risk Prediction Models: A Systematic Review; (4) Telomere Dynamics in Cancer: Novel Insights and Therapeutic Implications; and (5) Ubiquitinomics in Sigmoid Colon Cancer: Defining Stage-Specific Patterns for Predictive, Preventive, and Personalized Medicine. These chapters present new perspectives and a comprehensive overview of recent developments in molecular medicine.

However, it must be recognized that this book addresses only a tiny fraction of the field of molecular medicine, which aims to encourage researchers to contribute to molecular medicine to protect human health.

The target audience comprises scholars and specialists in the field.

Xianquan Zhan, MD, PhD, FRSM, FWAPS
Professor of Cancer Proteomics and PPPM,
Shandong Cancer Hospital and Institute,
School of Pharmaceutical Sciences,
Shandong First Medical University,
Qingdao Road, Jinan,
Shandong, China

Chapter 1

Immune-Related Gene Alternative Splicing-Based Biomarker-Sets for Lung Adenocarcinoma

Wenhao Sun, Na Li and Xianquan Zhan

Abstract

Lung adenocarcinoma (LUAD) poses a serious threat to human health, lacking effective biomarkers. Alternative splicing (AS) of immune-related genes (IRGs) is of great significance in LUAD. In this study, IRG-AS was analyzed in depth, and the prognostic model of IRG differential expression alternative splicing was constructed, and its relationship with tumor microenvironment, immune cells, and drug sensitivity was analyzed. The study obtained data from 513 LUAD tissues and 59 control tissues from the TCGA database and identified 2520 IRG-AS events, with exon skipping (ES) being one of the most common events. Through Lasso regression analysis, a prognostic model containing 12 IRG-DEAS events was established, which showed good risk stratification ability in both the training and test sets, with an area under the ROC curve of 0.84 and 0.676, respectively. In addition, the study also found that IRG-DEAS is closely related to the infiltration level of a variety of immune cells, immune checkpoint gene expression, and tumor mutational burden (TMB), which provides a new perspective for the prediction, prevention, and personalized medicine (3P medicine) of LUAD and a new idea for the precision treatment of lung cancer.

Keywords: lung adenocarcinoma, immune-related genes, alternative splicing, biomarkers, exon skipping, risk stratification, tumor mutational burden, prediction, prevention and personalized medicine (3 PM)

1. Introduction

1.1 Epidemiology of LUAD

Among the diseases that pose a serious threat to human health, lung cancer stands out with its high morbidity and mortality rates across the world [1]. The survey indicates that, around the world, lung cancer is one of the major reasons for cancer-related deaths. Each year, as many as 1.6 million people die from lung cancer and 1.8 million are diagnosed with the disease. Because lung cancers have different types, which could be diagnosed in clinical practice, the 5-year survival rate varies, which could be distributed from 4 to 17% [2]. Lung cancer includes three main kinds; of these, non-small cell lung cancer (NSCLC) is the most common type, growing relatively slowly; the other type is small cell lung cancer (SCLC), which is highly malignant; while lung carcinoids are relatively rare. One of the NSCLC subtypes is LUAD. Among all malignant lung cancers, nearly half of patients are diagnosed with LUAD. Its incidence rate is extremely high and can occur worldwide.

Many normal approaches are used to treat lung cancer. These methods develop rapidly, including chemotherapy, targeted therapy, surgery, and radiation. Apart from these, the most promising treatment is combination therapy [3]. However, it was still poor at the rate of survivors of lung adenocarcinoma, especially in some late-stage patients [4, 5]. A number of genetic mutations were studied to act as the factors driving lung cancer, such as ALK fusion, KRAS mutations, and EGFR mutations, and the use of some targeted drugs has achieved satisfactory results [6]. However, there were lots of patients diagnosed with lung cancer without a driven gene mutation. It is urgent that targeted drugs for these patients be developed. Therefore, it is necessary to identify new targets as the direction for drug treatment.

1.2 Alternative splicing (AS) in LUAD

Alternative splicing, a selective splicing process, transforms mRNA precursors into mature mRNAs. The generation of protein diversity depends on the fact that multiple mature mRNAs can be produced by a single gene. Scientists found that in the different cancers, a key mechanism linked to the biochemical mechanism was the change of AS; for instance, cancer cell proliferation, angiogenesis, and immune escape are facilitated by numerous cancer-related AS changes [7]. AS changes can make cancer cells come from the healthy cells, and this mechanism has the ability to function on all genes, leaving the frequency of oncogene mutations unaffected [8]. Due to alterations in AS patterns, the function or expression levels of lung cancer-associated genes might be altered, thereby influencing the occurrence and development of LUAD.

Alternative splicing events frequently occur in cancers and are intricately linked to tumor initiation, development, and treatment response. AS is acknowledged as a vital mechanism through which the protein diversity is boosted by the body, and the body can also regulate protein expression. However, the effective roles of AS in LUAD remain relatively underexplored [9]. Alternative splicing events within the immune process have attracted significant interest. For instance, immune-related genes [10] are prominently involved in numerous pathways. These pathways encompass chemokines and their receptors, B-cell antigen receptor (BCR) signaling cascade, TCR-mediated signal transduction pathway, antigen handling and presenting, interferon receptors, interleukin cytokines and their receptors, natural killer cell cytotoxicity, as well as members of the growth-regulating factor family of the transforming type and their corresponding receptors [11]. The alternative splicing of immune molecules generates novel splicing variants. These variants have the potential to influence how immune cells recognize and present antigens; then the activation of T cells will be impacted, and the initiation of the immune response will also be affected. Such an impact can, in turn, further modulate the efficacy of immunotherapy, as previously reported [12]. Transcriptomics-based investigations have shown that analyzing the AS patterns of IRGs in LUAD can help elucidate the underlying mechanisms through which AS affects immunotherapy. Additionally, this analysis can offer insights into the distinct functions of IRG isoforms within lung cancer contexts.

Dr. Zhan proposed the hypothesis that alternative splicing events of immune-related genes exhibit significant differences between lung cancers and control samples [13]. According to this hypothesis, lung cancer-specific AS events could be intricately linked to the molecular mechanisms underlying the disease, potentially serving as novel therapeutic targets. Moreover, specific AS patterns might act as accurate predictive or prognostic biomarkers within the precision preventive, predictive, and personalized medicine framework. To explore these hypotheses, a comprehensive analysis conducted by Dr. Zhan and his team detected the lung adenocarcinoma using data from the TCGA database [13].

In the analysis of lung adenocarcinoma, a total of 513 LUAD tissues from clinical patients and another 59 control tissues were employed for the study of immune-related gene alternative splicing (IRG-AS). As detailed in a research [13], 2520 IRG-AS events were identified. Among these, exon skipping (ES) emerged as one of the most prevalent alternative splicing events, whereas intron retention (ME) was the least common. An UpSet plot was utilized to illustrate the overlapping patterns among the seven types of IRG differentially expressed alternative splicing (IRG-DEAS) events in the LUAD group.

1.3 Immune-related genes (IRGs) and tumor microenvironment (TME)

Alternative splicing (AS) events of immune-related genes (IRGs) are intricately intertwined with the progression of tumors and the outcomes of new immunotherapy. Ample evidence has demonstrated that the tumor microenvironment (TME) undergoes distinct alterations in the context of AS, which is orchestrated by splicing factors (SF). These researches provide some new AS events, which can potentially modulate the efficacy of immunotherapy by influencing the expression levels of therapeutic targets.

Novel splicing variants, which are generated by the AS of immune molecules, have the potential to disrupt the processes by which immune cells recognize and present antigens. This disruption, in turn, impacts the initiation of the immunoreaction and the activation of $CD3^+$ cells. Ultimately, these cascading effects can have a profound influence on the overall effectiveness of immunotherapy.

2. Molecular landscape of LUAD

2.1 Genetic and epigenetic features

Weimin Li and his colleagues inferred copy number variations (CNV) by using inferCNV, according to the average expression of 100 genes from every karyon region [14]. They found that large-scale amplifications appear in patients with LUAD, mainly in the type II alveolar cells of the patients. For example, genes such as CD74, RPS14, GPX3, and TNIP1 on chromosome 5; CSNK1D, CD7, CYBC1, NARF, FPXK2, and TBCD on chromosome 17; as well as MED16, CFD, PTBP1, and others on chromosome 19 were amplified [14]. Subsequently, they computed the expansion state of genes, which may have the key clinical value, including EGFR, KRAS, BRAF, ERBB2, and MET. It was found that EGFR had obvious amplification peaks in subpopulations of clone 2 and subpopulations of clone 7 of the AT2 cells, and a great part of the cells in these subclones were from LUAD patients. After constructing the transcription factor governance network according to the upregulated genes, it was discovered that in LUAD, the tumor suppressor NKX2-1 functioned as one kind of cytokine of the critical regulators of AT2 cells, which are lineage-specific. Additionally, S100A13 was regulated by NKX2-1. In LUAD, it was found that around the transcription initiation point, the ATAC-seq peaks of S100A13 were higher than those in benign solitary pulmonary nodules (BSPN), and it was also higher in malignant AT2 cells in expression. Finally, through GO and DOSE analysis of the activated genes in aggressive cells, it was found that genes associated with the protein that is the unfolded response, cells that are related to immune, and mRNA metabolic regulation were enriched in AT2 cells of LUAD processes [14].

2.2 Tumor microenvironment and immune evasion

In developing studies, there has been a growing trend [15] in the prognostic splicing models by uncovering the system of alternative splicing and establishing correlations between splicing abnormalities and the tumor microenvironment as well as immune heterogeneity in lung adenocarcinoma [16]. As an illustration, a recent study identified a new AS-based prognostic signature related to Bregs and malignant pleural effusion in LUAD patients.

Additionally, a Swiss-led multicenter study has mapped the pan-cancer alternative splicing landscape *via* extensive analysis of 8705 patient tumors. Among 32 cancer types [17], lung adenocarcinoma emerges as a prominently AS-deregulated tumor, characterized by increased frequency of abnormal alternative splicing events. Notably, exon skipping (ES) events show an increase of more than 30% in tumor samples when compared to paired normal tissues [17]. In terms of therapeutic relevance, the study reveals that LUAD features a high frequency of tumor-specific exon-exon junctions [18]. This indicates a more intricate splicing mechanism and the potential for neoantigen production [19].

Remarkably, recent research efforts have developed spatial isoform transcriptomics (SITs) [20], an innovative approach designed to profile spatial alternative splicing (AS) isoforms and sequence heterogeneity *via* long-read sequencing techniques [21]. This novel SITs method enables *in situ* capture of full-length transcripts from diverse tissue regions, providing an AS landscape and comprehensive overview of both gene expression patterns simultaneously.

Significantly, in LUAD, the integration of spatial sequencing with RNA-seq has yielded crucial insights. It has been discovered that they are present at higher levels in cancer-related fibrocytes within cancer tissues [22, 23], such as hMENA/hMENAΔv6 splicing isoforms and so on. Additionally, stromal FN1 exhibits high expression, and ectopic lymphoid structures (TLSs) are primarily situated in the circumtumoral area. The abundant expression of FN1 might act as one kind of hindrance for the localization of intratumor TLSs, which is consistent with robust anti-tumor immune responses [22, 23]. Moreover, it may impact the effectiveness of immunotherapy by interacting with ILT3.

Furthermore, a new spliced type of PD-L1, known as sPD-L1, is also devoid of the membrane-spanning region, which has been detected in LUAD. The importance of using immune checkpoint blockade (ICB) monotherapy as pretreatment for LUAD patients with the combination of sPD-1 and sPD-L1 has been highlighted [24]. Interestingly, research results indicate that alternative splicing can produce an lncRNA from the PD-L1 gene. By increasing the activity of c-Myc, this lncRNA can accelerate the proliferation of LUAD [25, 26]. Overall, the latest findings underscore the complex functions of the AS of PD-L1 and its ligand in the proliferation of the cancer and immune modulation. These discoveries open up possible strategies if we want to make immunotherapy more effective for LUAD patients.

Immunotherapy represents a major breakthrough in the treatment of cancer, which uses the role of the immune system to detect and remove the tumor cells, including those in LUAD. Among the prominent approaches in this field is CAR-T cell adoptive cell therapy [27]. As an antibody-based anti-tumor therapy, it is considered one of the most promising personalized immunotherapeutic treatments [28, 29].

In LUAD, mRNA splicing differences were demonstrated to modify proteomic diversity, presenting a vast yet largely uninvestigated pool of potential novel immunotherapy targets. In immunotherapy target screening (IRIS), recent research emphasizes the value of a new substance from RNA splicing, which is named isoform peptides. IRIS employs an algorithmic framework engineered to identify tumor antigens arising from AS. This innovation significantly expands the horizons of cancer immunotherapies, offering new possibilities for more effective treatment strategies.

Among the emerging specific targets of the LUAD, CD44v6, which is the AS variant of CD44, has been associated with the development of the cancers, invasion, and cancer dissemination of LUAD [30]. *In vitro*, anti-CD44v6 CAR-T cells have revealed functional activation [31]. These CAR-T cells are competent of infiltrating tumors, proliferating, inhibiting *in vivo* LUAD growth, and improving overall survival in mouse models.

3. Immune-related gene alternative splicing (IRG-AS) in LUAD

3.1 Methodology for AS profiling

RNA sequencing has become a crucial methodological breakthrough, enabling in-depth, genome-wide investigations of differential gene expression and mRNA splicing patterns [2, 32]. One of the key strengths of this technology is single-cell RNA-seq. This technique enables the assessment of gene expression with a high precision of individual cells, allowing researchers to distinguish and characterize different cellular subsets within LUAD.

Moreover, the recent developments in spatial transcriptomics expand genomic analysis by maintaining the spatial organization of the LUAD structure [33, 34]. Interestingly, recent research has presented spatial isoform transcriptomics, an exploratory approach developed to describe spatial alternative splicing isoforms and sequence diversity using long-read sequencing. This novel SITs technique allows for the on-site capture of complete transcripts from various tissue regions. By doing this, it offers a holistic view of both the dynamics of gene expression and the landscapes of alternative splicing at the same time.

Scientists are actively working to overcome the inherent drawbacks of individual techniques and boost their ability to understand the biological roles of alternative splicing isoforms. Single-cell RNA sequencing with short-read assembly provides high sensitivity in detecting transcripts at the single-cell level. On the other hand, spatial transcriptomics allows the analysis of the patterns of expression of the genes within the tissue microenvironment. Overall, the integration of third-generation sequencing enriches the logical scope through revealing detailed information about AS events. In essence, a comprehensive strategy is offered by these advanced technologies for researching the translation of genes and the spatial characteristics of lung adenocarcinoma, thereby yielding a more in-depth understanding of how AS functions within the tumor microenvironment.

3.2 IRG-AS event characterization

After comparing the LUAD and normal tissues, an analysis of IRG-AS events identified 66 differential alternative splicing (DEAS) events in total, and we found

16 events were down-regulated and another 50 events were up-regulated [13]. KEGG pathway enrichment analysis was performed on IRG-related DEAS events, leading to the identification of 22 pathways in LUAD that have significance in terms of statistics (**Figure 1A**) [13] and multiple biological processes (**Figure 1B**).

Analyzing the identified DEASs by GO biological process enrichment, the study has uncovered 29 statistically significant biological processes in LUAD. A total of 127

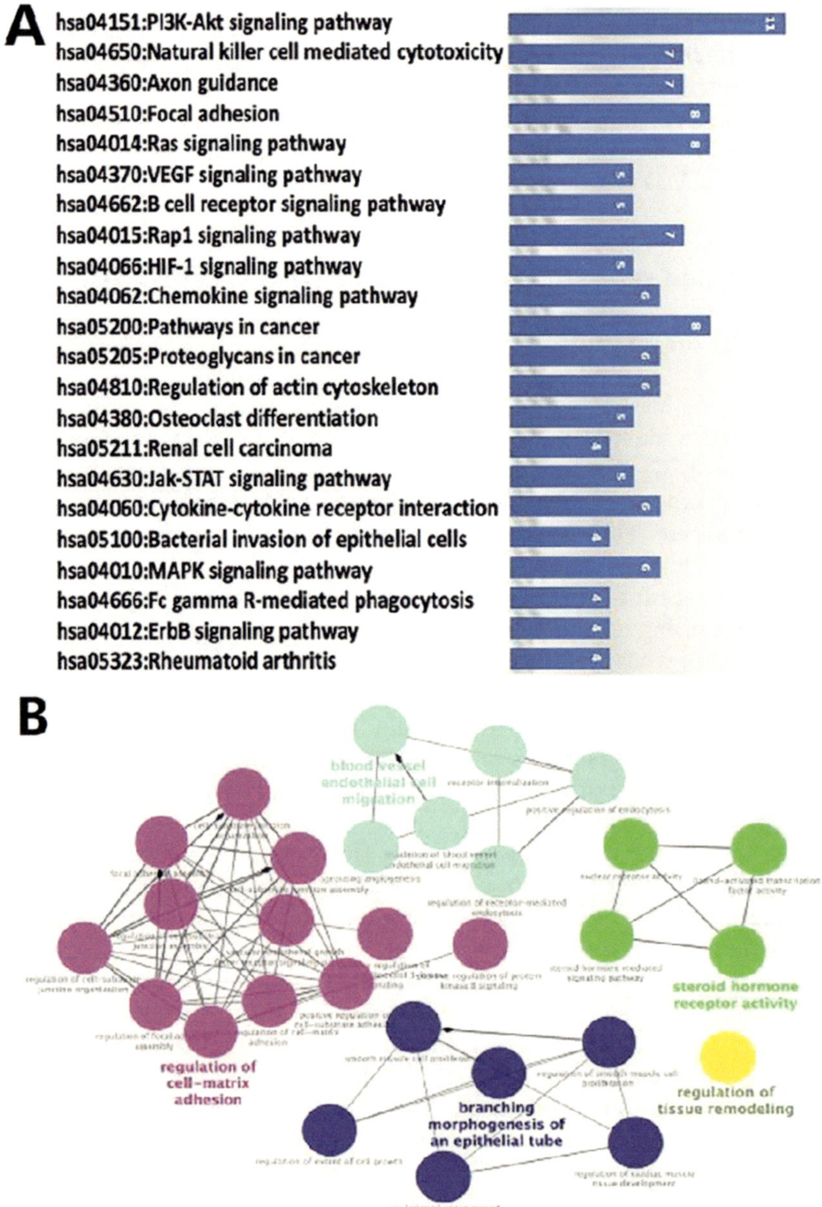

Figure 1.
The functional features of IRG-DEAS events in LUAD. A. The signaling pathways induced by IRG-DEAS were identified in LUAD by KEGG pathway enrichment analysis. B. IRG-DEAS-mediated biological processes (BPs) in LUAD. Reproduced from Li et al. [13].

immune-associated genes were subjected to variable splicing events to assess their association with overall survival in the LUAD cohort (**Figure 1B**) [13].

4. Prognostic model development and validation

4.1 Lasso regression and risk stratification

Between overall survival (OS)-related IRG-AS events and DEAS events, an overlapping analysis was carried out. Subsequently, the IRG-DEAS events, which are associated with OS, were subjected to the Lasso regression analysis. In the case of lung adenocarcinoma, 14 overlapping OS-associated IRG-DEAS events were identified, such as BMP1|82,995|AA and IL32|33,418|AA, among others (**Figure 2**) [13].

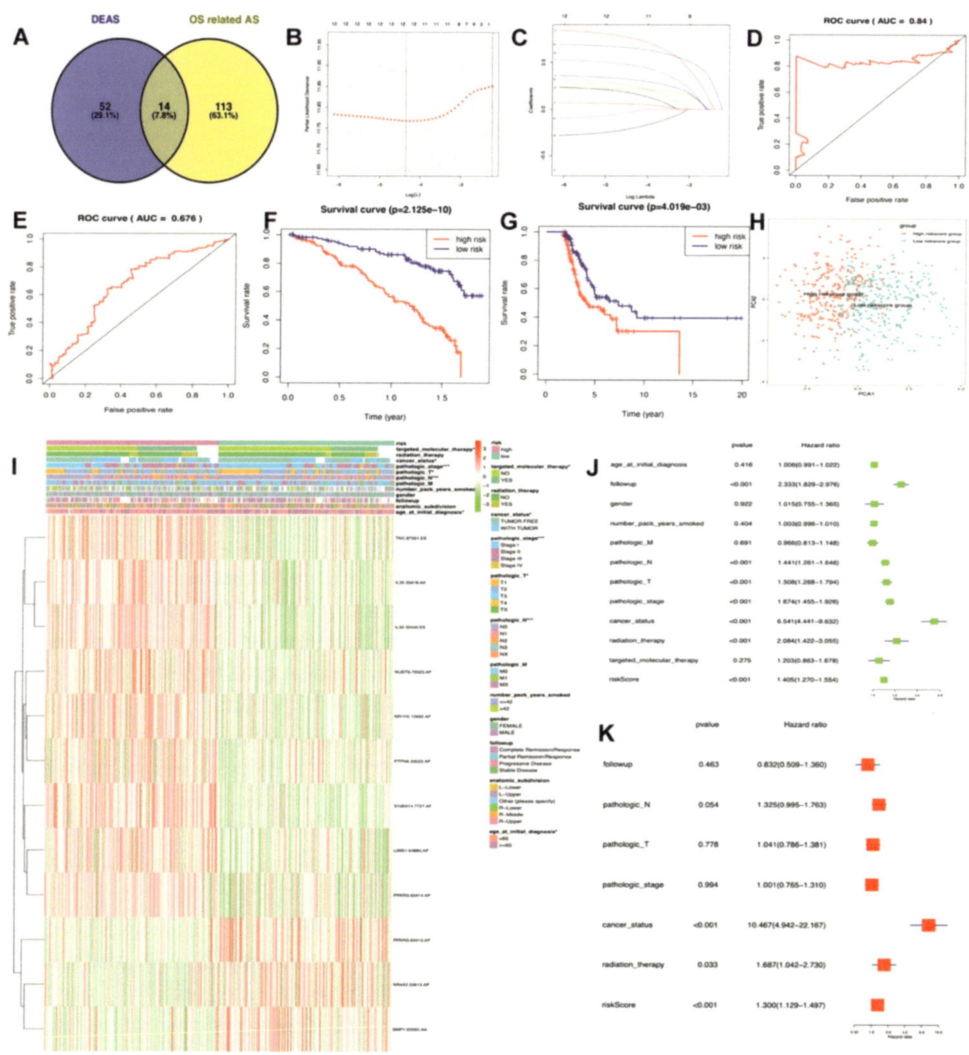

Figure 2.
Lasso regression identified the OS-related-IRG-DEAS prognostic model in LUAD. Reproduced from Li et al. [13].

When log(lambda) ranged between −5 and − 4, Lasso regression analysis applied to these 14 OS-associated IRG-DEAS events resulted in the identification of a 12-IRG-DEAS signature model. This model encompasses events such as IL32|33,418|AA, IL32|33,440|ES, and others [13].

4.2 Clinical and survival validation

According to the average risk score calculated using the model of 12-IRG-DEAS signature, LUAD tissue samples were classified into low- and high-risk score clusters using the IRG-DEAS signature in both training and testing cohorts [13]. ROC curve analysis demonstrated that the area under the curve values was 0.84 in the training set for LUAD and 0.676 in the testing set (**Figure 2D**). The analysis of overall survival (OS) showed that a notable statistical difference appeared in both training and testing datasets between the low-risk and the high-risk score groups (**Figure 2F**). Principal component analysis (PCA) revealed that all LUAD samples could be clearly classified into two distinct subgroups based on the risk scores evaluated for every sample according to the Lasso regression model (**Figure 2H**) [13]. Risk score, pathologic N, pT stage, tumor status, pStage, radiotherapy, and follow-up found by univariate analysis were significantly connected with OS. Multivariable statistics identified radiation therapy, risk score, and tumor condition as independent risk factors for LUAD (**Figure 2J,K**) [13].

5. Immune microenvironment and therapeutic implications

5.1 Immune cell dynamics

According to the 12-IRG-DEAS signature model in LUAD, notable differences were shown by nine types of immune cells between the low- and high-risk samples (**Figure 3B**). These immune cells were neutrophils, memory B lymphocytes, M0 macrophages, antibody-secreting cells, quiescent mast cells, mature dendritic cells, immature dendritic cells, monocytes, and effector CD4 memory T cells (**Figure 3A**) [13]. Subsequently, with a significance level established as $P < 0.05$, the correlations between the immune-related cells and risk score were reassessed in Corrplot software, which uses Spearman's correlation analysis. Among them, several immune cell types exhibited significant correlation coefficients. Among the cell types identified were CD8+ T cells, antibody-secreting cells, pro-inflammatory macrophages, naïve B cells, immature dendritic cells, monocytes, unpolarized macrophages, effector CD4+ memory T cells, resting mast cells, B memory cells, and neutrophils (**Figure 3C**) [13].

5.2 Immune checkpoints and therapy response

Several immune checkpoint genes showed notable differences between the low- and high-risk score samples. These genes included CD276, CD274, and CD80 (**Figure 3D**). The TIDE formula was utilized to estimate the propensity of response to immunotherapy among LUAD patients. A lower TIDE prediction score was revealed by the result, which pointed to low-risk score samples of LUAD patients (**Figure 3E**) [13].

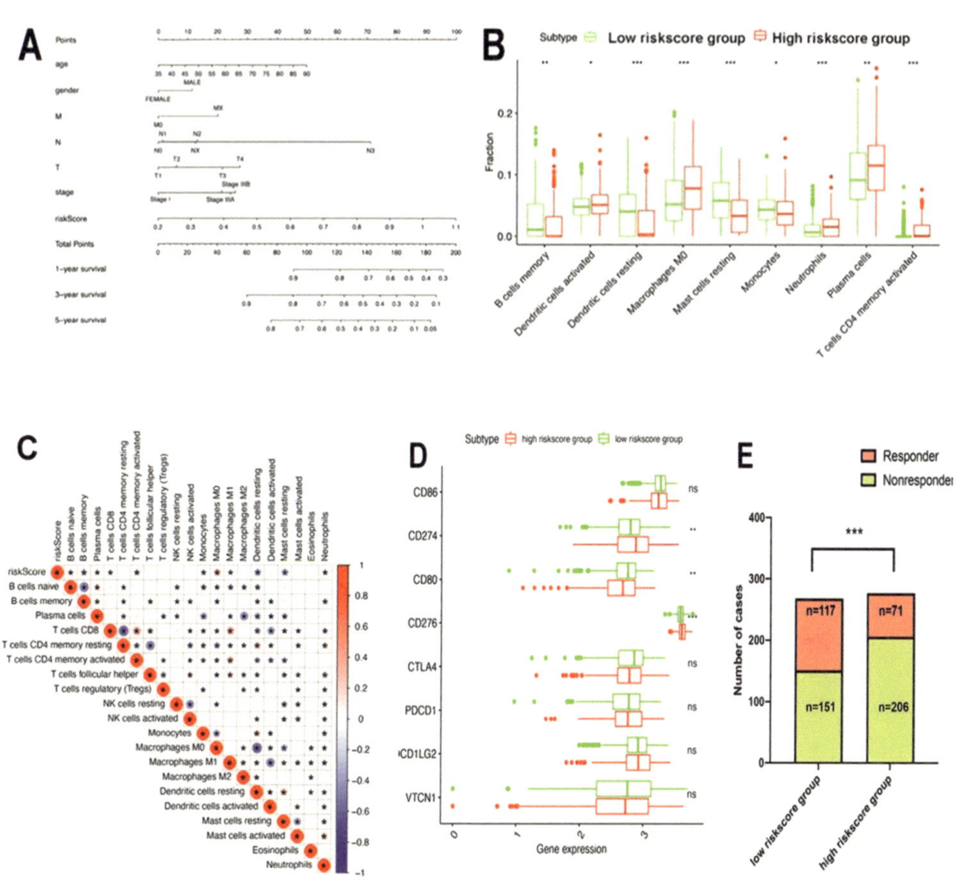

Figure 3.
In LUAD, immune status, drug susceptibility, and mutation status were analyzed between low- and high-risk scoring samples according to the prognostic model of OS-related IRG-DEAS characteristics. Reproduced from Li et al. [13].

5.3 Tumor mutation burden (TMB) and drug sensitivity

Between the TMB score and the risk score, analysis revealed a significant positive correlation (Cor = 0.2546, $P < 0.0001$) (**Figure 4B**) [13]. In LUAD, several IRG-AS occurrences identified by the Lasso model were significantly associated with medication sensitivity ($P < 0.05$). Certain IRG-AS occurrences exhibited a strong correlation coefficient. Examples include the associations between NR4A2 and mekinist, IL32 and DLS-10, IL32 and tanespimycin, as well as IL32 and tyrothricin (**Figure 4A**) [13].

6. Splicing factor (SF) regulatory networks

6.1 SF-AS interaction networks

The list of SF genes was utilized to query the SpliceAid database for the purpose of selecting genes. Based on a defined cutoff value, Cytoscape software was used to

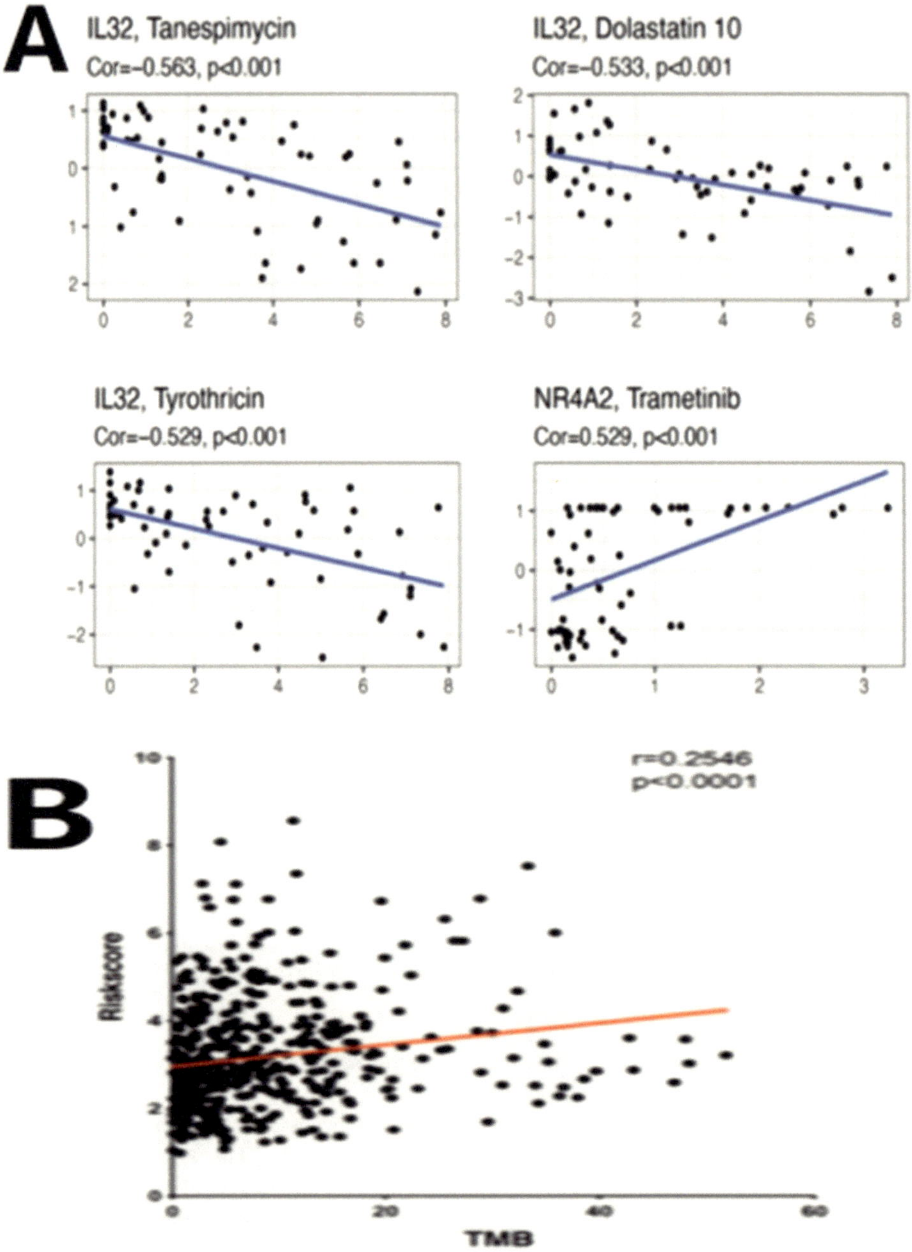

Figure 4.
A The association between identified gene expressions in the Lasso regression model and drug sensitivity in LUAD. B The association between risk scores and TMB scores in LUAD. Reproduced from Li et al. [13].

construct the system, which connects OS-associated IRG-AS events and SFs (correlation index>0.4: P < 0.001). The SF CELF6 was further investigated in LUAD cells by integrating the co-expression network of IRG-AS events, which was associated with overall survival and SFs in LUAD (correlation index >0.4: P < 0.001), with an analysis based on literature [13].

6.2 Functional validation of CELF6

A549 and H1299 were maintained in RPMI-1640 or DMEM added to 10% FBS, which are human LUAD cell lines obtained from ATCC. Si-control and si-CELF6 were synthesized, and the cells were incubated for 24-48 h after lipofecting with si-control or another using Lipofectamine 3000. Thereupon, multiple assays were performed: a wound healing assay for cell mobility, a CCK8 assay for cell proliferation, an EdU assay with fluorescence staining for positive proliferative cells, a Transwell chamber for cell invasion, an annexin V-FITC kit for apoptosis, and qRT-PCR was performed to validate the CELF6-targeted molecules within the SFs-OS-associated AS regulatory network. Trizol reagent was employed to extract total RNA from three independent biological samples of both the si-CELF6-treated and control samples. Finally, several IRG-AS events, including IRF3-51,008-ES, TNFRSF12A-33349-AD, etc., were verified between si-CELF6 and control groups [13].

Thereupon, si-CELF6 was transiently transfected into LUAD cells to conduct loss-of-function experiments. The CCK8 assay demonstrated that the multiplication of A549 and H1299 LUAD cell lines was enhanced after silencing CELF6. Specific qPCR validation results matched the analytical outcomes from the TCGA database analysis after using the SF-AS regulatory system. In A549 cells, when comparing the si-CELF6 group and the control group, this consistency was observed for TPM2-98, TYMP-62853-AA, 133-ES, TNFRSF12A-33349-AD, and BMP1-82,995-AA [13]. In H1299 cells, the matching results were found for TNFRSF12A-33349-AD and 008-ES, TPM2-98, and IRF3-51,133-ES between the control groups and si-CELF6 groups. Moreover, EdU, wound healing, and Transwell analyses demonstrated that for A549 and H1299 cells, the down-regulation of CELF6 significantly enhanced the growth and migration. Conversely, the knockdown of CELF6 notably reduced the apoptosis rate [13].

7. Clinical translation and 3P medicine

7.1 Innovations in 3P medicine

Currently, some clinical trials have been conducted for LUAD and have achieved favorable results (**Table 1**). LUAD can be treated through different methods. One such approach is stereotactic body radiotherapy (SBRT), which is capable of accurately delivering high-dose radiation to tumor sites while restricting the radiation exposure of adjacent healthy organs. This allows for ablative irradiation of the tumor, resulting in a high rate of local tumor control. Such a feature endows SBRT with potential utility in LUAD treatment. It serves as an effective local therapy option for LUAD patients who are ineligible for surgery due to advanced age, multiple coexisting conditions, or other factors. Furthermore, thanks to its precise radiation delivery, SBRT can minimize damage to vital organs like the surrounding normal lung tissue, heart, and esophagus during high-dose tumor irradiation. In doing so, it lowers the likelihood of complications such as radiation-induced pneumonitis and esophagitis, thereby enhancing patients' quality of life. This study still holds innovative significance for the 3P medicine in LUAD.

Predictive approach: Leveraging AS pattern differences between cancer and normal cells, reliable IRG-DEAS models serve as diagnostic markers and therapeutic targets. Combined with clinical data, these models enable hierarchical prediction of cancer risk and prognosis, advancing precision diagnostics in LUAD.

Trial code	Trial name	Drug/therapy	Phase	Patient population	Efficacy results	Reference
NCT02142738	KEYNOTE-024	Pembrolizumab (PD-1 inhibitor)	Phase III	Treatment-naïve advanced LUAD, PD-L1 expression ≥50%, without EGFR/ALK mutations	Median PFS: 10.3 months vs. 6.0 months (HR = 0.50, P < 0.001); Objective Response Rate (ORR): 44.8% vs. 27.8%.	[35]
NCT06068049	FLAURA	Osimertinib (3rd-generation EGFR-TKI)	Phase III	Treatment-naïve advanced LUAD, EGFR-sensitive mutations (19del or L858R)	Median PFS: 18.9 months (osimertinib) vs. 10.2 months (1st-gen TKI: gefitinib/erlotinib) (HR = 0.46, P < 0.0001).	[36]
NCT02409342	IMpower110	Atezolizumab (PD-L1 inhibitor)	Phase III	Treatment-naïve advanced LUAD, PD-L1-high (TC ≥ 50% or IC ≥ 10%), without EGFR/ALK mutations	Median OS: 20.2 months (atezolizumab) vs. 13.1 months (chemotherapy) (HR = 0.59, P = 0.01); ORR: 38.3% vs. 28.6%; Median Duration of Response (DoR): 38.9 months vs. 5.8 months.	[37]
NCT02075840	ALEX	Alectinib (ALK inhibitor)	Phase III	Treatment-naïve advanced LUAD, ALK fusion-positive	Median PFS: NA months (alectinib) vs. 11.1 months (crizotinib) (HR = 0.47, P < 0.001); ORR: 82.9% vs. 75.5%.	[38]
NCT02366143	IMpower150	Atezolizumab + bevacizumab + chemotherapy	Phase III	Treatment-naïve advanced non-squamous NSCLC (including LUAD), including EGFR/ALK-mutant patients	Quadruple therapy group median OS: 19.2 months vs. bevacizumab + chemotherapy group 14.7 months (HR = 0.78, P = 0.016); ORR: 63.5% vs. 48.0%; benefit also shown in EGFR-mutant subgroup (median OS 29.4 vs. 21.6 months).	[39]
NCT03215706	CheckMate 9LA	Nivolumab + ipilimumab + chemotherapy	Phase III	Treatment-naïve advanced NSCLC (including LUAD), regardless of PD-L1 expression or driver mutations	Dual immunotherapy + chemotherapy group median OS: 15.6 months vs. chemotherapy alone 10.9 months (HR = 0.66, P < 0.001); ORR: 38% vs. 25%; benefit retained in PD-L1-negative patients (HR = 0.65).	[40]
NCT02511106	ADAURA	Osimertinib	Phase III	Completely resected stage IB-IIIA LUAD, EGFR-sensitive mutations	2-year DFS rate: 90% (osimertinib) vs. 46.1% (placebo) in stage II-IIIA patients; significantly reduced risk of recurrence or death (HR = 0.17, P < 0.001); benefit also in stage IB patients (HR = 0.21).	[41]

Table 1.
Clinical trials targeting LUAD. The data is derived from ClinicalTrials.gov.

Targeted prevention: Splicing variants of immune molecules offer novel targets for immunotherapy, potentially enhancing specificity and reducing toxicity. Regulating AS of key genes or developing drugs against splicing variants could inhibit tumor growth and modulate immune responses, providing mechanistic insights for preventive strategies.

Personalized medical services: Multi-index models using AS events and clinical variables (e.g., stage, mutation status) enable patient stratification, improving prognosis accuracy and tailoring therapies. The established prognostic models effectively classify lung cancer subtypes, facilitating individualized treatment within the 3P framework [42].

7.2 Targeted therapeutic strategies

Certain splicing variants represent alternative forms of immunotherapy-targeted drugs and have the potential to exert more potent effects. One of the remedial targets and biomarkers to predict patient prognosis is IRG-AS. Therapies that target specific AS events may offer enhanced specificity and reduced side effects. CELF6, a protein that binds to RNA, acts as an antioncogene in multiple types of cancers. Focusing on CELF6 or other SFs could emerge as a novel therapeutic approach [42].

7.3 Challenges and future directions

Numerous hurdles persist within this domain despite the substantial potential of AS in lung cancer. Firstly, AS events in LUAD are highly intricate and fluid, necessitating in-depth investigations across multiple levels and time points. Secondly, substantial technical hurdles must be surmounted to effectively translate laboratory discoveries into practical clinical applications. Additionally, a profound understanding of the interactions between AS profiles and other biological features in LUAD, including genetic mutations and epigenetic modifications, is essential. For LUAD, it is advisable to incorporate AS profiles into the framework of proteoformics research. The developed OS-associated IRG-DEAS hallmark predictive model also needs validation with a larger human lung cancer sample cohort. This validation is crucial for developing reliable prognostic biomarkers that can facilitate patient stratification and personalized treatment for individuals with LUAD [35].

8. Conclusion

This study has comprehensively explored the role of IRG-AS events in LUAD, uncovering significant findings with far-reaching implications for both understanding the disease and developing novel therapeutic strategies.

Our research identified 66 DEAS events between LUAD and normal tissues, and through KEGG pathway and GO biological process enrichment analyses, revealed 18 significant pathways and 29 biological processes associated with these DEAS events. Using univariate Cox regression, we found 127 IRG-related AS events associated with OS in LUAD patients. Leveraging Lasso regression, we developed a 12-IRG-DEAS signature model, which demonstrated excellent performance in risk stratification, as validated by ROC curve analysis with AUC values of 0.84 in the training sample and 0.676 in the testing sample. Low- and high-risk samples were classified by this pattern effectively from LUAD patients, with notable discrepancy in the OS between the two groups, and for LUAD, multivariate analysis identified that one of the autonomous risk factors is the risk score.

This study also delved into the immune microenvironment of LUAD, showing that the 12-IRG-DEAS signature model was associated with significant differences in the large amount of multiple immune cell populations, checkpoint immune genes, and the likelihood of response to immunotherapy. Notably, between the risk score and TMB, there is a notable direct relationship, and several IRG-AS events were notably correlated with drug sensitivity, suggesting potential applications in predicting treatment response and guiding personalized therapy.

Furthermore, we constructed an SF-AS interaction network and identified CELF6 as a key SF for further study. Functional experiments demonstrated that CELF6 knockdown promoted the growth, migration, and aggression of LUAD cells while reducing apoptosis, highlighting its tumor-suppressive role in LUAD.

In terms of clinical translation, our nomogram analysis integrating the risk index and clinical features showed promise in predicting patient survival. The 12-IRG-DEAS signature model, along with specific IRG-AS events and SFs like CELF6, represents potential biomarkers and therapeutic targets that could enhance treatment specificity and reduce side effects.

However, several challenges remain. The complexity and dynamic nature of AS events in LUAD require more in-depth, multilevel, and longitudinal investigations. To effectively apply experimental findings into clinical practice, technical barriers must be surmounted. Additionally, understanding the interactions between the AS model and other biological features in LUAD is crucial. Incorporating AS events into proteoformics research and validating our prognostic model with larger cohorts are essential steps for developing reliable biomarkers and personalized treatment strategies for LUAD patients. In general, for future studies, the study provides a solid basis and paves the way to improve the prognosis of LUAD patients for novel therapeutic approaches.

Acknowledgements

This work was supported by the Shandong Provincial Taishan Scholar Engineering Project Special Funds (NO. tstp20221143 to X. Z.), the Shandong Provincial Natural Science Foundation (ZR2021MH156; ZR2022QH112), the Shandong First Medical University Talent Introduction Funds (to X. Z.), the Shandong First Medical University High-level Scientific Research Achievement Cultivation Funding Program (to X. Z.), and China National Nature Scientific Funds (81272798; 82203592).

Author's contributions

W.S. was responsible for formal analysis, visualization, and writing-original draft; N. L. was responsible for visualization and conceptualization. X. Z. was responsible for conceptualization, methodology, project administration, funding acquisition, writing-review and editing, and critically revision.

Conflict of interest

We declare that we do not have competing interests.

Author details

Wenhao Sun[1,2], Na Li[1] and Xianquan Zhan[1,2*]

1 Shandong Provincial Key Laboratory of Precision Oncology, Shandong Cancer Hospital and Institute, Shandong First Medical University and Shandong Academy of Medical Sciences, Jinan, Shandong, China

2 Shandong Provincial Key Medical and Health Laboratory of Ovarian Cancer Multiomics, and Jinan Key Laboratory of Cancer Multiomics, Medical Science and Technology Innovation Center, Shandong First Medical University and Shandong Academy of Medical Sciences, Jinan, Shandong, China

*Address all correspondence to: yjzhan2011@gmail.com

References

[1] Peralta-Arrieta I, Armas-López L, Zúñiga J, Ávila-Moreno F. Epigenetics in non-small cell lung carcinomas. Salud Pública de México. 2019;**61**(3):318-328. DOI: 10.21149/10089

[2] Nooreldeen R, Bach H. Current and future development in lung cancer diagnosis. International Journal of Molecular Sciences. 2021;**22**(16):8661. DOI: 10.3390/ijms22168661

[3] Nasim F, Sabath BF, Eapen GA. Lung cancer. Medical Clinics of North America. 2019;**103**(3):463-473. DOI: 10.1016/j.mcna.2018.12.006

[4] Lim JU, Yeo CD, Rhee CK, et al. Comparison of clinical characteristics and overall survival between spirometrically diagnosed chronic obstructive pulmonary disease (COPD) and non-COPD never-smoking stage I-IV non-small cell lung cancer patients. International Journal of Chronic Obstructive Pulmonary Disease. 2019;**14**:929-938. DOI: 10.2147/COPD.S190244

[5] Lee DY, Salahuddin T, Iqbal J. Lysine-specific demethylase 1 (LSD1)-mediated epigenetic modification of immunogenicity and immunomodulatory effects in breast cancers. Current Oncology. 2023;**30**(2):2127-2143. DOI: 10.3390/curroncol30020164

[6] Sgambato A, Casaluce F, Maione P, Gridelli C. Targeted therapies in non-small cell lung cancer: A focus on ALK/ROS1 tyrosine kinase inhibitors. Expert Review of Anticancer Therapy. 2018;**18**(1):71-80. DOI: 10.1080/14737140.2018.1412260

[7] Coomer AO, Black F, Greystoke A, Munkley J, Elliott DJ. Alternative splicing in lung cancer. Biochimica et Biophysica Acta (BBA) - Gene Regulatory Mechanisms. 2019;**1862**(11):194388. DOI: 10.1016/j.bbagrm.2019.05.006

[8] Jin M, Liu B, Chen C, et al. Genome-wide splicing quantitative expression locus analysis identifies causal risk variants for non-small cell lung cancer. Cancer Research. 2023;**83**(10):1742-1756. DOI: 10.1158/0008-5472.CAN-22-3184

[9] Ule J, Blencowe BJ. Alternative splicing regulatory networks: Functions, mechanisms, and evolution. Molecular Cell. 2019;**76**(2):329-345. DOI: 10.1016/j.molcel.2019.09.017

[10] Li N, Wang J, Zhan X. Identification of immune-related gene signatures in lung adenocarcinoma and lung squamous cell carcinoma. Frontiers in Immunology. 2021;**12**:752643. DOI: 10.3389/fimmu.2021.752643

[11] Bhattacharya S, Andorf S, Gomes L, et al. ImmPort: Disseminating data to the public for the future of immunology. Immunologic Research. 2014;**58**(2-3):234-239. DOI: 10.1007/s12026-014-8516-1

[12] Banerjee S, Galarza-Muñoz G, Garcia-Blanco MA. Role of RNA alternative splicing in T cell function and disease. Genes (Basel). 2023;**14**(10):1896. DOI: 10.3390/genes14101896

[13] Li N, Jia W, Wang J, et al. Clinically relevant immune subtypes based on alternative splicing landscape of immune-related genes for lung cancer advanced PPPM approach. The EPMA Journal. 2024;**15**(2):345-373. DOI: 10.1007/s13167-024-00366-4

[14] Zhang L, Zhang Y, Wang C, et al. Integrated single-cell RNA sequencing

analysis reveals distinct cellular and transcriptional modules associated with survival in lung cancer. Signal Transduction and Targeted Therapy. 2022;**7**(1):9. DOI: 10.1038/s41392-021-00824-9

[15] Whibley A, Kelley JL, Narum SR. The changing face of genome assemblies: Guidance on achieving high-quality reference genomes. Molecular Ecology Resources. 2021;**21**(3):641-652. DOI: 10.1111/1755-0998.13312

[16] Chen H, Gao F, He M, et al. Long-read RNA sequencing identifies alternative splice variants in hepatocellular carcinoma and tumor-specific isoforms. Hepatology. 2019;**70**(3):1011-1025. DOI: 10.1002/hep.30500

[17] Kahles A, Lehmann KV, Toussaint NC, et al. Comprehensive analysis of alternative splicing across Tumors from 8,705 patients. Cancer Cell. 2018;**34**(2):211-224.e6. DOI: 10.1016/j.ccell.2018.07.001

[18] Pan Y, Kadash-Edmondson KE, Wang R, et al. RNA dysregulation: An expanding source of cancer immunotherapy targets. Trends in Pharmacological Sciences. 2021;**42**(4):268-282. DOI: 10.1016/j.tips.2021.01.006

[19] Pan Y, Phillips JW, Zhang BD, et al. IRIS: Discovery of cancer immunotherapy targets arising from pre-mRNA alternative splicing. Proceedings of the National Academy of Sciences of the United States of America. 2023;**120**(21):e2221116120. DOI: 10.1073/pnas.2221116120

[20] Belchikov N, Hsu J, Li XJ, et al. Understanding isoform expression by pairing long-read sequencing with single-cell and spatial transcriptomics. Genome Research. 2024;**34**(11):1735-1746. DOI: 10.1101/gr.279640.124

[21] Lebrigand K, Bergenstråhle J, Thrane K, et al. The spatial landscape of gene expression isoforms in tissue sections. Nucleic Acids Research. 2023;**51**(8):e47. DOI: 10.1093/nar/gkad169

[22] Paavola KJ, Roda JM, Lin VY, et al. The fibronectin-ILT3 interaction functions as a stromal checkpoint that suppresses myeloid cells. Cancer Immunology Research. 2021;**9**(11):1283-1297. DOI: 10.1158/2326-6066.CIR-21-0240

[23] Di Modugno F, Di Carlo A, Spada S, et al. Tumoral and stromal hMENA isoforms impact tertiary lymphoid structure localization in lung cancer and predict immune checkpoint blockade response in patients with cancer. eBioMedicine. 2024;**101**:105003. DOI: 10.1016/j.ebiom.2024.105003

[24] Ruiz-Cordero R, Devine WP. Targeted therapy and checkpoint immunotherapy in lung cancer. Surgical Pathology Clinics. 2020;**13**(1):17-33. DOI: 10.1016/j.path.2019.11.002

[25] Qu S, Jiao Z, Lu G, et al. PD-L1 lncRNA splice isoform promotes lung adenocarcinoma progression via enhancing c-Myc activity. Genome Biology. 2021;**22**(1):104. DOI: 10.1186/s13059-021-02331-0

[26] Nielsen C, Ohm-Laursen L, Barington T, Husby S, Lillevang ST. Alternative splice variants of the human PD-1 gene. Cellular Immunology. 2005;**235**(2):109-116. DOI: 10.1016/j.cellimm.2005.07.007

[27] Hong M, Clubb JD, Chen YY. Engineering CAR-T cells for next-generation cancer therapy. Cancer Cell.

2020;**38**(4):473-488. DOI: 10.1016/j.ccell.2020.07.005

[28] Neelapu SS, Locke FL, Bartlett NL, et al. Axicabtagene Ciloleucel CAR T-cell therapy in refractory large B-cell lymphoma. The New England Journal of Medicine. 2017;**377**(26):2531-2544. DOI: 10.1056/NEJMoa1707447

[29] Maalej KM, Merhi M, Inchakalody VP, et al. CAR-cell therapy in the era of solid tumor treatment: Current challenges and emerging therapeutic advances. Molecular Cancer. 2023;**22**(1):20. DOI: 10.1186/s12943-023-01723-z

[30] Porcellini S, Asperti C, Corna S, et al. CAR T cells redirected to CD44v6 control tumor growth in lung and ovary adenocarcinoma bearing mice. Frontiers in Immunology. 2020;**11**:99. DOI: 10.3389/fimmu.2020.00099

[31] Yoshida C, Kadota K, Yamada K, et al. CD44v6 downregulation as a prognostic factor for distant recurrence in resected stage I lung adenocarcinomas. Clinical and Experimental Medicine. 2023;**23**(8):5191-5200. DOI: 10.1007/s10238-023-01185-z

[32] Li N, Zhan X. Signaling pathway network alterations in human ovarian cancers identified with quantitative mitochondrial proteomics. The EPMA Journal. 2019;**10**(2):153-172. DOI: 10.1007/s13167-019-00170-5

[33] Li N, Zhan X. Identification of clinical trait-related lncRNA and mRNA biomarkers with weighted gene co-expression network analysis as useful tool for personalized medicine in ovarian cancer. The EPMA Journal. 2019;**10**(3):273-290. DOI: 10.1007/s13167-019-00175-0

[34] Zaman S, Hajiran A, Coba GA, et al. Aberrant epidermal growth factor receptor RNA splice products are among the Most frequent somatic alterations in clear cell renal cell carcinoma and are associated with a poor response to immunotherapy. European Urology Focus. 2021;**7**(2):373-380. DOI: 10.1016/j.euf.2019.12.001

[35] Reck M, Rodríguez-Abreu D, Robinson AG, et al. Pembrolizumab versus Chemotherapy for PD-L1–Positive Non–Small-Cell Lung Cancer. New England Journal of Medicine. 2016;**375**(19):1823-1833. DOI: 10.1056/NEJMoa1606774

[36] AstraZeneca. An Ambispective, Non-Interventional, Multiple Cohort Study to Assess the Management of Osimertinib Treatment in Patients with EGFRm Non-Small Cell Lung Cancer under Real-World Conditions in Spain. Spain: AstraZeneca Clinical Study Information Center; 2025. Available from: https://clinicaltrials.gov/study/NCT06068049 [Accessed: July 22, 2025]

[37] Herbst RS, Giaccone G, Marinis F de, et al. Atezolizumab for First-Line Treatment of PD-L1–Selected Patients with NSCLC. New England Journal of Medicine. 2020;**383**(14):1328-1339. DOI: 10.1056/NEJMoa1917346

[38] Roche H-L. Randomized, Multicenter, Phase III, Open-Label Study of Alectinib Versus Crizotinib in Treatment-Naive Anaplastic Lymphoma Kinase-Positive Advanced Non-Small Cell Lung Cancer. Switzerland: F. Hoffmann-La Roche, Ltd; 2025. Available from: https://clinicaltrials.gov/study/NCT02075840 [Accessed: July 22, 2025]

[39] Socinski MA, Jotte RM, Cappuzzo F, et al. Atezolizumab for first-line treatment of metastatic nonsquamous NSCLC. The New England Journal of Medicine.

2018;**378**(24):2288-2301. DOI: 10.1056/NEJMoa1716948

[40] Paz-Ares L, Ciuleanu T-E, Cobo M, et al. First-line nivolumab plus ipilimumab combined with two cycles of chemotherapy in patients with non-small-cell lung cancer (CheckMate 9LA): An international, randomised, open-label, phase 3 trial. The Lancet Oncology. 2021;**22**(2):198-211. DOI: 10.1016/S1470-2045(20)30641-0

[41] Roy S Herbst, Yi-Long Wu, et al. Osimertinib in EGFR-Mutated Lung Cancer. New England Journal of Medicine. 2021;**384**(7):675-676. DOI: 10.1056/NEJMc2033951

[42] Bajinka O, Ouedraogo SY, Golubnitschaja O, Li N, Zhan X. Energy metabolism as the hub of advanced non-small cell lung cancer management: A comprehensive view in the framework of predictive, preventive, and personalized medicine. The EPMA Journal. 2024;**15**(2):289-319. DOI: 10.1007/s13167-024-00357-5

Chapter 2

Exploring the Multifaceted Role of Wnt Signaling: From Stemness, Development, and Disease to Therapeutics

Nima Ghasemi, Hossein Azizi and Thomas Skutella

Abstract

Signaling pathways form intricate networks that govern essential cellular processes, including differentiation, proliferation, and tissue homeostasis. These pathways enable intercellular and intracellular communication through molecular mediators such as proteins and genes. Among the highly conserved signaling pathways, such as Notch, Hedgehog, TGF-β/BMP, and Hippo, Wnt signaling plays a pivotal role in embryonic development, stem cell proliferation, self-renewal, and tissue regeneration. This pathway, governed by the Wnt gene family, operates through canonical (β-catenin-dependent) and non-canonical (β-catenin-independent) mechanisms. Both canonical and non-canonical Wnt/β-catenin pathways regulate a wide range of normal and pathological biological processes. Wnt proteins act as secreted morphogens and are pivotal in embryonic development, stem cell function, tissue repair, fate determination, and adult tissue remodeling. However, dysregulation of Wnt signaling has been implicated in various diseases, including cancer. This chapter explores different facets of this biological pathway across various stem cell types, including embryonic, hematopoietic, mesenchymal, induced pluripotent, epidermal, and cancer stem cells. It examines the pathway's functional roles, gene expression dynamics, and protein-protein interaction (PPI) networks through an in silico approach. Moreover, Wnt/β-catenin signaling is a focal point in clinical contexts, stem cell therapy, and regenerative medicine, a topic to be addressed herein. By analyzing its biomarkers, therapeutic targets, and emerging research directions, this review underscores the Wnt signaling pathway's complexity and its transformative potential in both fundamental biology and translational medicine.

Keywords: Wnt signaling, Wnt/β-catenin pathway, PPI network, gene expression, stemness, cancer, therapeutics

1. Introduction

Signaling pathways orchestrate events within and outside cells, mediated by molecular factors like molecules, proteins, and their adjacent genes. These pathways intersect and communicate with each other through crosstalk, forming complex signaling networks. While numerous signaling pathways are known, further research is essential for a deeper understanding of this field. These pathways and networks regulate cellular processes such as differentiation, proliferation, and specialized functions. Stem cells, with their remarkable ability to differentiate into various cell types and undergo self-renewal, hold immense potential for clinical applications and regenerative medicine. Understanding their biological regulatory networks and signaling pathways is crucial for harnessing their therapeutic potential effectively. Additionally, cancer and malignant tumors are subjects of intense research, with a focus on understanding their cellular and biological properties, including their signaling networks. Stem cells and cancerous cells alike are influenced by various signaling pathways, including the pi3k-akt-mTOR, Notch, Hedgehog, and Wnt pathways [1]. Among these pathways, the Wnt signaling pathway has emerged as particularly significant, exerting profound effects on both stem cells and malignant cells [2, 3]. Its potential impact on various cellular processes makes it a promising target for future experiments and translational research in applied sciences. The Wnt signaling pathway plays crucial roles in various biological processes, including wound healing, stem cell maintenance, tissue homeostasis, and embryonic development [4]. The intricate signaling transduction cascade, recognized as the Wnt signaling pathway, is intricately governed by signaling molecules encoded by Wnt genes. It was first in 1982 in the study of virally-induced C3H murine breast tumors that Nusse and Varmus reported cloning of the integration-1 proto-oncogene [5]. Later, it was revealed that the fruit fly gene wingless (Wg), critical for segment polarity, is its homolog [6]. Subsequent investigations have revealed the broader scope of Int-1's significance, revealing its membership in a highly conserved gene family now recognized as the wingless-type mouse mammary tumor virus (MMTV) integration site (Wnt) gene family. The Wnt signaling pathway is separated into canonical (or β-catenin-dependent) and non-canonical (or β-catenin-independent) pathways, influencing signaling networks, each exerting profound effects on cellular signaling networks. The canonical pathway governs crucial aspects of cell fate, proliferation, and survival, while the non-canonical pathway modulates differentiation, cellular properties, and migration, collectively shaping the intricate landscape of cellular biology [7, 8]. In the canonical Wnt signaling pathway, the indispensable players include the frizzled (FZD) family, coreceptor alongside the low-density lipoprotein (LDL) receptor family, and LDL-related protein (LRP). Conversely, the non-canonical Wnt signaling pathway applies its influence through various channels, such as the modulation of frizzled family receptors, interactions with ROR2, RYK, small G proteins, and the Wnt/Ca2 + −dependent signaling pathway [7, 9]. These intricate mechanisms highlight the multifaceted nature of Wnt signaling, where different pathways collaborate to regulate diverse cellular processes with precision and complexity. A comprehensive understanding of the regulatory mechanisms governing these pathways and the signaling molecules involved helps us to use the potential of signaling pathways for clinical applications. Manipulating or modulating the Wnt signaling pathway has opened new doors in the field of medicine and regenerative therapies. For example, the utilization of the Wnt signaling pathway holds promise in lung regeneration and repair [10], cancer treatment [11–13], bone and skeletal disorders treatment [14], and glioblastoma treatment [15]. This review aims to delve into the wide array of Wnt signaling pathways, exploring their multifaceted roles in both normal and cancerous cells. We will examine Wnt signaling across various cellular groups, including embryonic stem cells, hematopoietic stem cells, mesenchymal stem cells, iPSCs (induced pluripotent stem cells), epidermal stem cells, and cancer stem cells. Furthermore, we will explore the clinical applications of Wnt signaling and their future perspective alongside an investigation into the PPI network from an in silico perspective. This comprehensive analysis will shed light on the intricate roles of Wnt signaling pathways and their potential implications for fundamental research and therapeutic interventions. Wnt proteins, classified as morphogens, affect cellular properties and movement by creating concentration gradients.

2. Wnt signaling pathway

The human WNT gene family comprises 19 distinct members, each playing unique roles in cellular signaling [16]. Upon activation, Wnt signals initiate signaling pathways that have diverse effects on cellular behavior. These Wnt signaling pathways are categorized into canonical and non-canonical pathways, each modulating distinct cellular responses such as cell fate determination, survival, and cell polarity [17–19]. This complicated network of Wnt signaling pathways shows the multifaceted nature of cellular regulation and highlights the importance of precise signaling modulation in governing various biological processes.

2.1 Wnt proteins

Wnt proteins are comprised of 19 secreted glycoproteins, each weighing approximately 40 kDa [20]. They feature an N-terminal hydrophobic signaling sequence and 22–24 conserved cysteine (Cys) residues. The formation of disulfide bonds among the conserved Cys residues within Wnt proteins, coupled with N-glycosylation and lipid modifications, assumes important roles in protein modification and eliciting cellular responses. Wnt proteins undergo posttranslational modifications, including N-glycosylation, as well as unique lipid modifications (O-acylation) on conserved Cys residues, highlighting their distinct characteristics as secreted proteins [21–24]. The lipid modifications observed involve the attachment of palmitic acid to specific cysteine residues and palmitoleic acid to serine residues [24–26]. These modifications influence structural stability, membrane targeting, and protein-protein interactions [26, 27]. Notably, lipid modifications are not exclusive to Wnt proteins; they have also been observed in Hedgehog and Drosophila EGFR ligand Spitz [26, 28]. Despite extensive research, the precise role of lipid modification in Wnt signaling has not been fully understood. While the loss of the palmitate moiety affects Wnt signaling activity, its exact function needs further studies. However, research has provided evidence indicating that lipid modifications, notably palmitoyl acyl groups, play a crucial role in Wnt secretion. Specifically, alterations on Cysteine93 and Serine239 have been identified as essential for facilitating proper Wnt secretion. Intriguingly, investigations have revealed that single mutants of Wnt proteins at either Cysteine93 or Serine239 have normal secretion patterns. But, when mutations occur at both sites simultaneously, there is a notable decrease in Wnt secretion [7]. Recent findings suggest a direct interaction between lipid-modified Wnt and the frizzled receptor, highlighting the potential signaling mechanisms [29, 30]. Posttranslational modification of Wnt proteins starts with entering the endoplasmic reticulum (ER). A multipass transmembrane ER protein known as porcupine (Porc), which belongs to the membrane-bound O-acyltransferase (MBOAT) family, plays a significant role in modifications of Wnt ligands [31]. Studies have demonstrated that Porc-deficient cells cannot secrete Wg, and the loss of Porc leads to a decrease in the hydrophobicity of Wg [32]. Furthermore, it has been shown that the increase in the hydrophilicity of mWnt3 is accompanied by overexpression of Porc [21].

The secretory pathways begin after posttranslational modification of Wnt proteins, which in this path, move from the endoplasmic reticulum (ER) to the trans-Golgi network (TGN) and ultimately to various subcellular compartments [33].

In the extracellular matrix (ECM), the presence of Wnt proteins involves a crucial interaction with heparan sulfate proteoglycans (HSPGs), which are pivotal in preventing their aggregation within the ECM environment. This interaction stabilizes the Wnt proteins, ensuring their solubility within the aqueous conditions of the ECM. This mechanism is essential not only for preserving the structural integrity of the Wnt proteins but also for facilitating their proper functioning within the intricate cellular milieu [34]. To ensure the proper delivery of Wnt ligands to their destination, the Wnt secretory pathway relies on proteins. One such specialized protein is the conserved multipass transmembrane protein, known as Wntless/Evi/Sprinter (Wls), which serves as a key effector in the Wg secretory pathway, facilitating the transport of Wnt ligands from the TGN to the plasma membrane [35, 36]. In Drosophila, its absence results in the Wg molecules secretion failure, while other secreted proteins such as Hedgehog (Hh) remain unaffected [37, 38]. Additionally, in *Caenorhabditis elegans*, it has been discovered that the retromer complex is present in Wnt-producing cells and is involved in intracellular Wnt trafficking [39]. However, there are conflicting results about the involvement of the retromer complex in Wnt-secreting cells. While some studies have shown that Wnt signaling remains unaffected in retromer-deficient cells [40], others have found that these cells have blocked Wnt secretion [41]. This deficiency in Wnt secretion was retrieved by Wls overexpression [42]. Further analysis has revealed that the retromer complex and Wls bind to each other, showing a close relationship between Wls and the retromer complexes in the Wnt signaling pathway [43]. These data may indicate that the retromer complex is required for a specific Wnt secretion that needs further investigation in future research.

In the Drosophila wing imaginal disc, Wg affects gene transcription so that high concentrations of Wnt morphogens induce gene transcription near Wnt-producing cells, while lower concentrations activate genes across the wing area [44–46]. Despite being a hydrophilic molecule because of lipid modifications, Wnt molecules move across the extracellular spaces to exert their effects [47]. Various models have been proposed to explain this mechanism. Passive diffusion, planar transcytosis, and the "bucket brigade" relay mechanism are unlikely routes for Wnt diffusion due to the hydrophobic properties of Wnt molecules and the long estimated time required for Wnt molecules to cross between cells in this model [34, 47, 48]. The most acceptable method for Wnt molecule diffusion through the extracellular matrix is carrier-assisted diffusion, wherein Wnt molecules bind to a diffusible molecule like Swim to facilitate easier movement of Wnt molecules in the aquatic extracellular environment over long distances and prevent the Wnt molecules aggregation [49, 50]. These findings show how Wnts apply their effects despite being hydrophobic and offer insights into the mechanisms of morphogen distribution.

2.2 Canonical Wnt signaling pathway: Β-catenin dependent pathway

The canonical Wnt signaling pathway is considered a well-known pathway among various Wnt signaling pathways. At the center of this pathway is β-catenin, regulating gene transcription post-transcriptionally as a co-activator of the Wnt signaling cascade.

Wnt proteins such as Wnt2b, Wnt3, Wnt3a, Wnt8a, Wnt8b, Wnt9a, and Wnt10b are known as participants in the canonical Wnt signaling pathways [51]. In the absence of Wnt proteins, the "destruction complex" maintains a low level of β-catenin within the cell. This complex comprises casein kinase 1α (CK1α), glycogen synthase kinase 3 β (GSK-3β), adenomatous polyposis coli (APC), and axis inhibition (Axin) [52]. The frizzled (Fzd) family proteins are characterized by ~7 transmembrane G protein-coupled receptors (GPCRs) and possess a cysteine-rich N-terminal end with around ten cysteine residues. Wnt ligands interact with Fzd receptors *via* their N-terminal cysteine-rich domain (CRD) [53]. Without the Wnt ligands, both frizzled and LRP receptors are inactive, resulting in the sequestration of β-catenin proteins in the destruction complex with the effect of APC and Axin [54]. Within this complex, β-catenin undergoes phosphorylation at serine (Ser) 45, Ser33, Ser37, and threonine 41 (Thr41) by the two key components of the destruction complex, casein kinase 1α CK1α and GSK-3β. This phosphorylation event creates a binding site for the β-transduction repeat-containing protein (β-TrCP) E3 ubiquitin ligase [54], promoting the recruitment of protein subunits Cul1/Cdc53 and Skp1, thereby forming the Skp1-Cullin-F-box (SCF) E3 ubiquitin ligase complex, SCF-β-TRCP. Subsequently, ubiquitination targets β-catenin for proteasomal degradation [55]. Consequently, this mechanism effectively inhibits gene transcription by preventing β-catenin's nuclear translocation, thereby allowing gene repressors to maintain their histone deacetylation and chromatin compaction functions [56]. Activation of the canonical Wnt signaling pathway occurs upon binding of Wnt molecules to FZD family receptors and LRP5/6 coreceptors, triggering LRP5/6 co-receptor phosphorylation [57]. Downstream interactions in Wnt signaling are propelled by the cytoplasmic protein Disheveled (DVL) [57]. Three homologs of the DVL protein have been identified, each featuring three conserved domains in both humans and mice. These domains facilitate distinct protein-protein interactions, thereby initiating specific branches of the Wnt signaling pathway [7, 58]. Following the LRP5/6 coreceptors phosphorylation, DVL undergoes phosphorylation upon activation of the Wnt-frizzled-LRP5/6 complex. This phosphorylation event results in the DVL recruitment to the plasma membrane, effectively inhibiting the destruction complex [57]. Consequently, β-catenin proteins are released from the destruction complex, where their phosphorylation by GSK-3β is inhibited, preventing their degradation *via* the proteasomal system and leading to an increase in cytoplasmic β-catenin levels [31]. Among the various binding partners of β-catenin that regulate gene transcription, the T-cell factor (TCF)/lymphoid enhancer factor (LEF) transcription factor family is critical. Groucho (Gro) binds to specific gene promoter sites and acts as a gene transcription suppressor (**Figure 1**) [59]. However, the presence of intranuclear β-catenin leads to the dissociation of Gro and its replacement with β-catenin molecules. This replacement, along with the partnership of β-catenin and TCF/LEF factors, allows chromatin remodeling and activates β-catenin target genes [57]. This gene transcription activation can be mediated by the recruitment of transcriptional co-activators and histone modifiers, including the ATP-dependent helicase Brahma-related gene 1 (BRG1), cyclic adenosine monophosphate response element (CREB)-binding protein (CBP), p300, B-cell lymphoma9 (BCL9), and pygopus [60].

2.3 Non-canonical Wnt signaling pathway: B-catenin independent pathway

The non-canonical Wnt signaling pathway, although it is less well-known than its canonical Wnt signaling pathway, plays crucial roles in cellular processes, with emerging roles in carcinogenesis. This pathway comprises two main extensively studied branches: the Wnt/PCP and Wnt/Ca2+ signaling pathways. The Wnt/PCP pathway

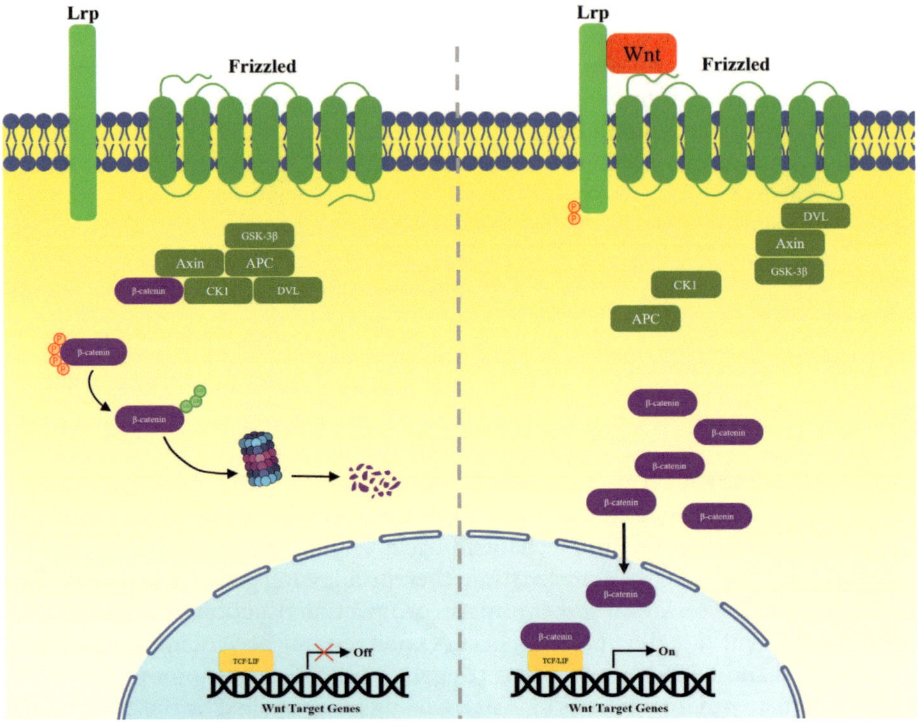

Figure 1.
Wnt/β-catenin signaling pathway in the absence and presence of Wnt ligand. The left panel shows the "Off" state of the Wnt/β-catenin pathway, where the absence of Wnt ligand leads to the degradation of β-catenin. In this state, β-catenin is phosphorylated by a destruction complex consisting of GSK-3β, CK1, Axin, and APC. Phosphorylated β-catenin is ubiquitinated and targeted for proteasomal degradation, preventing β-catenin from translocating to the nucleus and activating target gene transcription. The right panel illustrates the "On" state of the pathway upon Wnt ligand binding to frizzled and Lrp receptors. This binding disrupts the destruction complex, allowing β-catenin to accumulate in the cytoplasm and translocate into the nucleus, where it associates with TCF/LEF transcription factors to activate Wnt target gene expression.

affects actin polymerization, exerting broad effects on DAAM/RhoA/ROCK signaling. Conversely, the Wnt/Ca2+ pathway leads to increased intercellular calcium levels, activating downstream signaling molecules such as calcium/calmodulin-dependent protein kinase II (CAMKII), protein kinase C (PKC), and calcineurin. These events will eventually result in the activation of various effectors, including the nuclear factor of activated T-cells (NFAT) as a target of calcineurin, and CDC42 as a target of PKC, influencing gene transcription and cytoskeletal rearrangement. Additionally, both pathways also modulate cell polarity and migration. Notable Wnt ligands involved in non-canonical signaling pathways are Wnt4, Wnt5a, Wnt5b, Wnt7a, Wnt7b, and Wnt11 [61]. It is crucial to note that the same Wnt ligands such as Wnt3a, Wnt4, Wnt5a, Wnt5b, Wnt7b, and Wnt16 may have the potential to activate both canonical and non-canonical Wnt signaling pathways. This dual functionality shows the complicated and context-dependent nature of Wnt signaling, where the outcome can vary based on cellular conditions and interactions with specific receptors and coreceptors [62]. Wnt ligands exert their influence by binding to frizzled receptors, as well as coreceptors such as RAR-related orphan receptor (ROR) 1/2, receptor-like tyrosine kinase (Ryk), and protein tyrosine kinase 7 (PTK7) coreceptors.

The Wnt/PCP signaling pathway plays a pivotal role in various cellular processes, including cytoskeletal rearrangement, cell motility, cell polarity, morphogenesis, and migration [62]. This pathway is particularly crucial during critical developmental events such as gastrulation, hair follicle organization, and the orientation of stereocilia in the inner ear's sensory epithelium [63, 64]. Notably, the planar cell polarity (PCP) regulated pathway can influence cellular polarity in both epithelial and non-epithelial contexts [65]. In this signaling pathway, Wnt binds to FZD and ROR2 receptors, sending a signal to DVL, which leads to the formation of the Disheveled-associated activator of morphogenesis (DVL-Daam-1) complex. This complex activates small GTPases such as ras homolog gene family member A (RhoA), RHO, and RHO-associated kinase (ROCK) [54]. Additionally, DVL itself can activate ras-related C3 botulinum toxin substrate (RAC), JUN-N-terminal kinase (JNK), and the activator protein-1 (AP-1), and these activated effectors, including RHO, JNK, and ROCK, exert control over gene transcription and activate target genes [66]. Two distinct signaling pathways emerge from this activation: one involving the DVL-Daam-1 complex, which regulates cytoskeletal rearrangement through actin polymerization and microtubule stabilization by affecting RhoA, RHO, and ROCK, and the other, which influences cell motility and polarity by Dvl effects on RAC, AP-1, and JNK [67]. This complex interaction of signaling molecules and effectors shows the multifaceted nature of the Wnt/PCP pathway in shaping cellular behavior and tissue morphogenesis.

The Wnt/Ca^{2+} signaling pathway plays critical roles in cell fate determination, dorsal axis formation, tissue homeostasis, and cell migration [67]. In the Wnt/Ca^{2+} signaling pathway, Wnt proteins bind to frizzled receptors (FZD). This interaction between the Wnt ligand and frizzled, along with the phosphorylation of the co-receptor Ror by GSK3β, triggers phospholipase C (PLC) activation [68]. PLC activation leads to the transformation of phosphatidylinositol (4,5)-biphosphates (PIP2) to inositol (1,4,5)-triphosphates (IP3) and diacylglycerol (DAG). This conversion acts as a signal to the release of intracellular calcium from the ER, activating calcineurin (CaN) and calcium/calmodulin-dependent kinase II (CamKII). DAG and calcium ions released from the ER activate protein kinase C (PKC) [57]. CamKII and PKC activate NF-κβ and CREB regulatory factors. Additionally, Nuclear factor-activated T-cells (NFAT) are activated by CaN. These signaling molecules regulate gene transcription by translocating into the nucleus [31]. Besides, calmodulin activation triggers TGF-β-Activated kinase 1 (TAK-1) and Nemo-like kinase (NLK) activation, which function in opposition to canonical Wnt signaling pathways [31, 62]. While the non-canonical Wnt signaling pathway is known to play critical roles in cell regulation and gene transcription, its intricate details and alternative pathways warrant further investigation and analysis.

Understanding the complex and multifaceted signaling pathways, such as the Wnt signaling pathway, and their impact on cellular behaviors and regulatory mechanisms is crucial for further advances in research and clinical areas. By going deeper into these pathways and uncovering new insights comprehensively, researchers can better understand cellular processes and potentially identify novel targets for therapeutic applications. This knowledge can pave the way for breakthroughs in various fields, including regenerative medicine, cancer treatment, tissue engineering, and developmental biology. As we continue to explore and illuminate the complexities of cellular signaling, we open up new doors for innovation and discovery that promise to improve human health and well-being.

3. Wnt signaling pathway in different stem cells

Stem cells are a versatile group with the remarkable ability for self-renewal and differentiation into various cell types. They rely on complicated sets of regulatory mechanisms, including gene transcription, epigenetics, and various signaling pathways, to sustain their functions and activities such as self-renewal, stemness maintenance, and differentiation. The equilibrium between self-renewal and differentiation is crucial in maintaining various tissues within the body and ensuring their homeostasis. Among these pathways, the Wnt signaling pathway holds a significant role in the regulation and control of different stem cell populations across different cell lineages. Its involvement ensures the proper functioning and maintenance of stem cells, contributing to their resilience and adaptability in various physiological contexts.

3.1 Embryonic stem cells (ESCs)

Embryonic stem cells (ESCs) are pluripotent stem cells that are derived from the inner cell mass (ICM) of the blastocyst, capable of forming all three embryonic layers: ectoderm, mesoderm, and endoderm. Their decision to proliferate and self-renew or differentiate highly depends on intercellular and extracellular signals and factors. ESCs are used in *in vitro* studies to explore the pluripotency properties of stem cells. To prevent ESC differentiation *in vitro*, they are cultured on a layer of mouse embryonic fibroblast feeder (MEF) cells or in the presence of leukemia inhibitory factor (LIF). LIF prevents differentiation by promoting STAT3 phosphorylation through the activation of receptor tyrosine kinase gp130 [69, 70].

There are conflicting reports about the effect of the Wnt signaling pathway on ESCs and pluripotency. It has been found that the Wnt signaling pathway maintains the expression of transcription factors such as Nanog, Oct4, c-Myc, and Sox2, which promote pluripotency and inhibit differentiation [71]. H7 hESCs cells express RNA for all 19 Wnt genes and express all 10 Fz receptors, as well as both LRP5 and LRP6 coreceptors [72]. It has been found that knockout in GSK-3α, GSK3-3β, and APC encoding genes promotes pluripotency and limits differentiation by activating the Wnt signaling pathway [73]. It has been demonstrated that 6-bromoindirubin-3′-oxime (BIO), a GSK3-β inhibitor, can increase expression of Oct3/4, Rex1, and Nanog expression in ESCs. Additionally, it has been clarified that BIO is also able to activate STAT3. So, the Wnt signaling pathway can maintain ESCs pluripotency along with LIF/STAT3 pathway activation [74–76]. These findings gave us the capability to culture ESCs without any need for LIF or MEFF. *In vitro* studies have demonstrated that LIF and BMP4 cannot inhibit extracellular signal-related kinase (ERK), a factor that triggers differentiation in ESCs by mitogen-activated protein kinase/extracellular signal-related kinase (MAPK/ERK) [77]. However, inhibition of the MAPK/ERK signaling pathway by CHIR99021 small molecules can accompany GSK-3β inhibition without the need for LIF/STAT3, resulting in stemness maintenance in ESCs [77]. In addition to these mechanisms related to Wnt signaling pathways, Tcf3 binds to the same promoter sites as those occupied by specific pluripotency genes such as Oct4, Nanog, and Sox2 and inhibits gene expression. Activation of the Wnt signaling pathway and β-catenin relieves Tcf3-mediated repression on the pluripotency-related genes [78, 79].

In contrast, it has been demonstrated that in the self-renewal of hESCs, the Wnt/β-catenin signaling pathway is not active, and Wnt signaling pathway inhibition does not affect ESCs' self-renewal and Wnt signaling induction in hESCs results in promoting mesodermal-related genes [80, 81]. Further studies demonstrated that Wnt signaling pathway activation promotes differentiation in cooperation with other signaling pathways such as PI3K/Akt, Raf/Mek/Erk, and Smad2/3 [82].

Research has shown that in some cases the Wnt signaling pathway affects pluripotency in ESCs and helps them to remain in their pluripotency state and proper balance between self-renewal and differentiation. However, in contrast to this debate, other researchers have mentioned that the Wnt signaling pathway can affect differentiation and act as a differentiation promoter. These conflicting results show the need for further research in characterizing the specific role of the Wnt signaling pathway in ESCs. Digging more into the Wnt signaling pathway related to pluripotency control could help us to more advances in pluripotency-related research and achieve more and better pluripotent stem cells.

3.2 Mesenchymal stem cells (MSCs)

Mesenchymal stem cells (MSCs) are multipotent stem cells that express canonical and non-canonical Wnt signaling-related proteins such as Wnt ligands, receptors, and regulators. These signaling pathways are crucial for the processes of self-renewal, proliferation, and differentiation of MSCs [83].

The Wnt signaling pathway contributes to the differentiation of MSCs into adipocytes and the regulation of adipogenesis. MSCs differentiate into pre-adipocytes and then further differentiate into mature adipocytes. Within this differentiation process, CCAAT/enhancer binding protein (C/EBP) β and δ promote adipogenesis by inducing the expression of peroxisome proliferator-activated receptor γ (PPARγ) and C/EBPα [84]. Inhibition of the Wnt signaling pathway is a key driver of adipogenesis, with studies showing that overexpression of Wnt ligands, such as Wnt1, Wnt6, Wnt10a, and Wnt10b, inhibits adipogenesis [33]. The canonical Wnt/β-catenin pathway inhibits adipogenesis, where the β-catenin-TCF/LEF complex represses PPARγ expression, inhibiting pre-adipocyte differentiation into mature fat cells [33]. Wnt/β-catenin independent pathways can have a dual function; for instance, Wnt5a and Wnt5b upregulate PPARγ and promote adipogenesis [33] while activation of the CaMKII-TAK1-NLK pathway can accompany β-catenin-TCF/LEF by repressing PPARγ expression and blocking adipogenesis [85]. There is also evidence that shows Wnt signaling represses the differentiation of brown adipocytes, as demonstrated in the FAB4-Wnt10b mouse model [86]. Understanding the molecular mechanisms underlying adipogenesis, such as the Wnt signaling pathway, provides insights into the regulation of adipogenesis and potential therapeutic targets for conditions related to adipose tissue dysfunction, such as obesity.

The Wnt signaling pathway also plays a significant role in osteogenesis. Overexpression of Wnt10b leads to higher levels of osteoblast transcription factors (Runx2, Osx, and Dlx5), which result in higher bone mass and strength in mice, showing a positive effect of the Wnt signaling pathway on bone formation [87]. Additionally, loss-of-function mutants in LRP5 coreceptors are associated with osteoporosis-pseudo glioma [88], while gain-of-function mutations in LRP5 coreceptors are related to a high bone mass phenotype [89]. Non-canonical Wnt signaling pathways such as Wnt5a-ROR2 signaling also enhance osteogenesis by stimulating the expression of osteogenic transcription factors Runx and Osterix [90]. The CaMKII-TAK1-NLK signaling is effective in Runx overexpression in bone marrow MSCs by PPARγ repression [91]. However, some studies indicate that Wnt1-activated Wnt signaling can negatively impact osteogenesis by downregulating transcription factors such as Runx2, Dlx5, and Osterix [92]. Digging into the complex interplay between Wnt signaling pathways and osteogenesis provides insights into the molecular mechanisms underlying bone formation and offers potential therapeutic targets for conditions.

In the context of chondrogenesis, the Wnt signaling pathway appears to have a vital role. Activation of the Wnt1 signaling pathway inhibits SOX9 expression, a transcription factor necessary for chondrogenesis, and the progression through various chondrogenic stages. This suggests that Wnt signaling inactivation is essential for chondrogenesis and the maturation and maintenance of chondroid cells [93].

The Wnt signaling is crucial for developing and maintaining skeletal muscle. It regulates the proliferation and differentiation of satellite cells, which are essential for muscle repair and regeneration. Both canonical and non-canonical Wnt signaling pathways are involved in muscle regeneration [94]. Activation of canonical Wnt signaling in rat MSCs induces pro-myogenic factors like Pax7, MyoD, and myogenin while inhibiting adipogenic factors like C/EBPα and PPARγ expression [91]. Wnt7a, by binding to Fz7, promotes the expansion of satellite stem cells by activating the Wnt PCP pathway [94]. Additionally, in differentiated muscle cells, Wnt7a can activate the PI3K/Akt/mTOR growth pathway, linking the expansion of the satellite cell pool to muscle tissue mass [94]. A proper vision around the Wnt signaling pathway in muscle regeneration and development can lead us to novel therapeutic strategies for muscle-related disorders and injuries.

3.3 Germ stem cells (GSCs)

The Wnt signaling pathway is believed to be present in germ cells. Wnt/STOP signaling is crucial for promoting the first embryonic cell divisions in mature Xenopus oocytes [95]. This pathway stabilizes mitotic effectors such as Polo-like kinase 1 (Plk1) and ubiquitin-conjugating enzyme E2C (Ube2c) by inhibiting glycogen synthase kinase 3 (GSK3) [95]. Additionally, Wnt/LRP6 signaling is also essential for the maturation of transcriptionally silent sperm [96]. Ccnyl1 mutations, a component of Wnt/LRP6 signaling in mice are responsible for male-infertile mice, characterized by immotile sperm [97]. During their passage through the epididymis, testicular sperm receive Wnt signals that are critical for their terminal differentiation and acquisition of motility. Wnt2b secreted by the epididymal epithelium is involved in Wnt/LRP6 signaling, and overexpression of the Wnt inhibitor Dkk1 in the epididymis has been shown to reduce sperm motility, whereas truncated APC, which disrupts the APC function, increases it [98]. This understanding offers valuable insights into the signaling cascades regulating germ cells and highlights potential therapeutic targets for addressing male infertility.

3.4 Hematopoietic stem cells (HSCs)

Hematopoietic stem cells (HSCs) are responsible for the generation of all lineages of blood cells. Wnt signaling pathway among multiple signaling pathways, including Notch, Hedgehog, and Wnt, is believed to play crucial roles in promoting human HSCs multiplication and inhibiting their differentiation, suggesting that Wnt ligands act as hematopoietic growth factors. Overexpression of activated β-catenin expands HSCs in long-term cultures and triggers increased self-renewal, indicating the role of canonical Wnt signaling in HSC regulation [99]. Research indicates that a deficiency in Wnt3a leads to a reduction in both the number and reconstitution capacity of fetal liver HSCs, as well as impairs the self-renewal of bone marrow HSCs [100]. However, some counter-arguments propose that the impact of Wnt signaling on HSC function could be dependent on its dosage. Mild levels of Wnt activation enhance the repopulating capacity of HSCs, whereas excessive activation hinders HSC self-renewal. Intermediate levels of Wnt activation enhance myeloid differentiation potential, whereas high levels induce T-cell differentiation potential [101]. Additionally, the non-canonical Wnt4 ligand regulates HSCs and activates genes required for cell maintenance in mice [102]. Activation of Wnt signaling in HSCs increases the expression of HoxB4 and Notch1, genes known to be involved in HSC self-renewal [103]. Moreover, it has been proposed that Wnt signaling plays a role in establishing responsive chromatin architecture during HSC ontogeny, which is maintained in adults and is essential for HSC homeostasis [104]. Wnt signaling plays a complex role in regulating HSCs' fate decisions, and more studies are required to understand this path, which may have implications for therapeutic interventions targeting blood-related disorders.

3.5 Neural stem cells (NSCs)

Mouse models with mutated Wnt ligands such as Wnt1 and Wnt3a showed the Wnt signaling pathway as an important element in the development of various brain regions like the telencephalon, metencephalon, and diencephalon. This pathway is essential for the maintenance of neural stem cells (NSCs) in adult neural tissue. When β-catenin is activated, it boosts the proliferation of neural progenitors, whereas blocking the canonical Wnt pathway reduces neurogenesis [105]. The Ryk receptor forms a ternary complex with Fzd and Wnt1, resulting in Wnt-induced neurite outgrowth [31]. However, there are inconsistent findings regarding the impact of the β-catenin Wnt signaling pathway on neurogenesis. Genetic polymorphisms related to the Wnt pathway have been associated with psychiatric disorders like schizophrenia and autism in humans. Dysregulation of Wnt signaling, including increased expression of the Wnt antagonist Dkk1, is linked to the pathology of Alzheimer's disease, suggesting a role in synapse maintenance and plasticity [91]. The potential of the Wnt signaling pathway's impact on neurons and brain development is an interesting area for further investigation.

3.6 Mammary stem cells (MaSCs)

Research has found that a rare population of cells can re-form a functional mammary gland that can form mammary stem cells (MaSCs). Wnt signaling effectively affects the self-renewal of MaSCs and the normal development of mammary gland I under the influence of Lrp5 and Lrp6 canonical pathway coreceptors [106]. Wnt signaling in the mammary stem cell niche acts to inhibit differentiation by suppressing Gata-3, a transcription factor that promotes the differentiation of mammary stem cells into luminal cells [107].

3.7 Wnt signaling pathway in abnormal and cancerous stem cells

Cancer stem cells (CSCs) represent a subset of cells within tumors that exhibit characteristics of both stem cells and cancer cells. These cells possess the ability to self-renew and differentiate into various cell types present in the tumor. The expression patterns of transcriptional regulators in CSCs overlap with those in human ESCs, including genes involved in stemness and pluripotency such as NANOG and OCT3/4. This shared transcriptome influences the differentiation status and aggressiveness of tumors [108].

Aberrant activation of the Wnt signaling pathway has been implicated in various cancers, including colon, breast, hepatocellular, melanoma, ovarian, and lung cancers [109]. Studies have shown that the stabilization of β-catenin and activation of Wnt target genes as a result of loss-of-function mutations in the APC gene are associated with colorectal cancer (CRC) [109]. Besides, mutations in the Axin gene can disrupt β-catenin regulation and promote malignancy. Mutations in the β-catenin gene can result in its stabilization, leading to activation of Wnt signaling [110]. However, additional mutations in genes such as p53, PTEN, K-Ras, PIK3CA, and SMAD4 are typically required for full carcinogenesis [110]. Both canonical and non-canonical Wnt signaling pathways are believed to be associated with different cancer types and their stem cells. In prostate cancer, Wnt3a stimulates β-catenin and increases the sphere size of CSCs [111]. Wnt3a is also associated with glioblastoma and interacts with Pspo2, enhancing the undifferentiated state of CSCs. In Wnt3a-Rspo1-associated glioblastoma, this gene profile results in Lgr5 upregulation, which affects target stemness [112]. In non-canonical pathways, Wnt3a and Wnt5a stimulate sphere formation through β-catenin-independent transcriptional activity involving calcium mobilization. Inhibition of phospholipase C and NFAT impairs self-renewal [113]. Additionally, overexpression of Wnt5a increases CD44+ CD24- CSCs, enhancing chemoresistance *via* PKC phosphorylation [113, 114].

Frizzled (Fzd) receptors are believed to be involved in different cancer types and CSCs. Fzd7 interacts with Wnt5a/b, regulating Lgr5, CD44, Smad3, Yap1, and STAT3, and involves mammosphere formation in breast cancer [115]. Fzd6 is another receptor that is effective in neuroblastoma-derived cells, promoting neurosphere formation and invasiveness of malignant cells. Fzd6+ cells with the association of Wnt ligands express MYC, CD44, Cyclin D1, Twist, and TH genes, exhibiting chemoresistance and malignancy profiles [116].

CSCs are often resistant to conventional treatments such as chemotherapy and radiation therapy. They express multidrug resistance genes and can rapidly induce DNA repair mechanisms, making them resilient to treatment. Additionally, CSCs can stimulate angiogenesis, facilitating tumor development. Efforts are underway to develop alternative therapeutic strategies that selectively target CSCs, including antagonists of the Wnt/β-catenin pathway. By specifically targeting CSCs, researchers aim to overcome treatment resistance and improve outcomes for cancer patients.

3.8 Intestinal stem cells (ISCs)

The epithelium of the small intestine and colon contains crypts, which harbor ISCs at their base [117]. These ISCs give rise to progenitor cells capable of differentiating into various epithelial lineages, contributing to the continuous renewal of the intestinal epithelium [118]. Canonical Wnt signaling is therefore essential for maintaining ISC homeostasis and ensuring proper intestinal function. Research showed the important role of the Wnt signaling pathway in intestinal crypts and its influence on ISCs [119]. It has been demonstrated that inhibition of the Wnt signaling pathway results in complete ablation of intestinal crypts [120]. Additionally, the Wnt inhibitor DKK1 overexpression disrupts proliferation in adult intestinal crypts [121]. Wnt signaling affects ISCs and forms a gradient along the crypt-villus axis. This gradient controls cell fate decisions, with high Wnt and low BMP signaling promoting progenitor cell proliferation and low Wnt and high BMP signaling promoting differentiation [122]. Paneth cells in the base of the crypt produce essential growth factors, such as EGF and Wnt3a, which are necessary for ISC proliferation and preservation of stem cell properties [123]. Paneth cells regulate Lgr5-containing cells and maintain their maintenance by secreting Wnt3a as a signaling molecule [124]. Wnt3a is then transferred to Lgr5+ cells and retained by the frizzled receptor. This localization results in regulations of E3 ligases (Rnf43/Znrf3) and Wnt signaling agonists (Lgr4–5/R-Spondin) within the crypt [125]. Previous studies have revealed the involvement of β-Catenin and T-cell factor (TCF) in controlling the expression of cytoskeleton-modifying receptors EphB2/EphB3 and their ligand, ephrin-B1. This regulation allows for the coordination of cell proliferation, differentiation, and movement along the crypt axis in the intestine.

3.9 Hair follicle stem cells

The canonical Wnt signaling pathway is crucial for leading bulge stem cells toward the hair cell fate, rather than regulating epidermal differentiation [126]. Human hair follicle stem cells (HFSCs) reside in a niche within the hair follicle where they respond to various signaling ligands that regulate hair follicle development, maintenance, and regeneration after injury. The hair follicle undergoes cyclical phases of growth (anagen), regression (catagen), and rest (telogen), with Wnt/β-catenin signaling playing a pivotal role in each phase [91]. Upregulation of Wnt signaling at the end of telogen promotes the transition to anagen, facilitating hair growth [127]. However, during telogen, Wnt upregulation is suppressed by BMP signaling [128]. Lgr5 expression is observed in various locations within the hair follicle at different stages of the hair cycle. The progeny of Lgr5-positive cells repopulate stem cell compartments, supporting the existence of a stem or progenitor cell hierarchy within the hair follicle [129, 130]. In the context of hair follicle development, the activation of Lef1/Tcf complexes by beta-catenin leads to the upregulation of target genes, which leads to the formation and activation of new hair follicles.

3.10 Gastric stem cells

Wnt signaling is one of the key factors in the gastric stem cell niche environment. It regulates the self-renewal and differentiation of GSCs, similar to its role in intestinal stem cells. In neonatal mice, cells from all parts of the stomach epithelium originate from Lgr5-expressing cells, similar to the small intestine [131]. But, in adult mice, Lgr5 cells are primarily found in the pylorus [132]. Gastric stem cells are thought to reside at the bottom of the gland and produce differentiating progeny that migrate upward toward the gland. Like intestinal stem cells, GSCs rely on the presence of a niche environment for their maintenance and regulation [91]. In the stomach, Wnt signaling is crucial for the self-renewal and differentiation of GSCs. Specifically, the Wnt ligand Wnt3a appears to play a key role in regulating the function of GSCs, similar to its function in intestinal stem cells [91, 131]. More research is needed to highlight the importance of the Wnt signaling pathway in maintaining gastric tissue.

4. Protein-protein interactions (PPI) network of involved genes

To achieve the gene list involved in the Wnt signaling pathway, we used enrichr libraries and selected the KEGG 2021 Human library. The gene list consisted of 166 genes, and pathway figures were gathered from the Wnt signaling pathway (hsa04310) from KEGG (**Figure 2**) [133, 134].

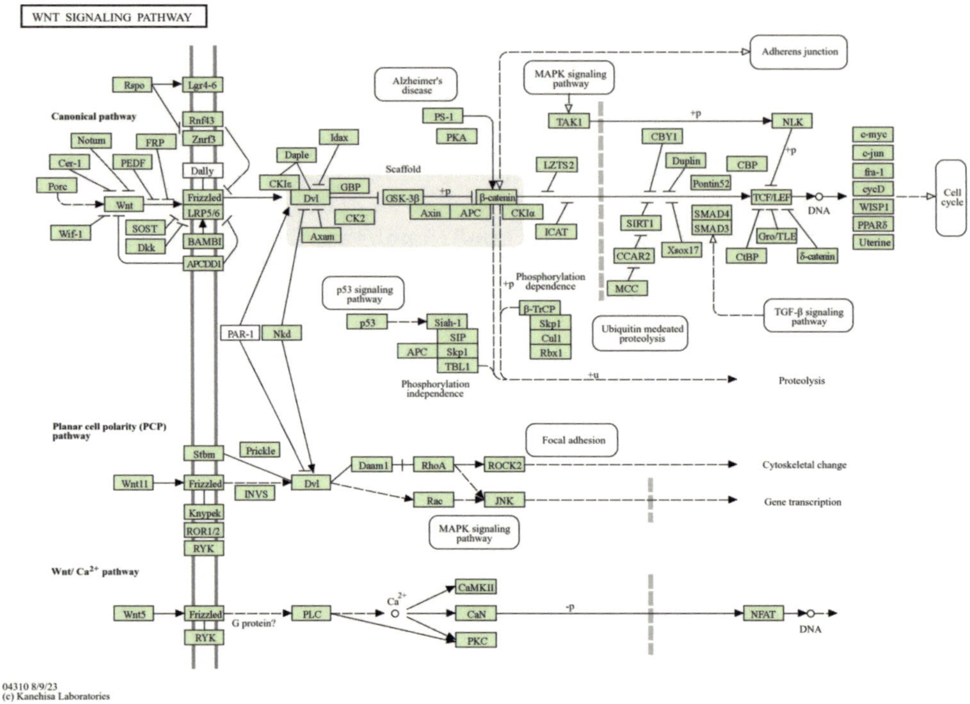

Figure 2.
The Overview of the Wnt signaling pathway and involved proteins generated by KEGG.

The Wnt signaling pathway gene list was then imported to STRING (v.12.0) for performing protein-protein interactions (PPI) network analysis and visualization based on the STRING database. The STRING analysis results involving the PPI network files were downloaded for further analysis. We imported the PPI network to the Cytoscape app (v.3.6.0) for network analysis and performed mathematical patterns such as degree, betweenness centrality, closeness centrality, and eigenvector centrality for filtering network nodes that give as important genes named "hub genes," which can give us a proper vision of the most central genes in the network that can be predicted to have more involvement in driving Wnt signaling pathway. Based on our analysis, GSK3β, AXIN1, MYC, PORCN, DKK1, CCND1, and LRP5 are just some of the highlighted hubs (**Figure 3**).

In the next step, for delving more deeply into gathered data and analysis results, we imported these hubs into the enrichr database and performed enrichment analysis. This enables us to dig into predicted pathways and look for any beneficial and significant pathway that can be used in future research. The enrichment analysis gathered from gene ontology (GO) revealed that neuron differentiation (GO: 0030182), cellular response to retinoic acid (GO:0071300), endocytic vesicle membrane (GO:0030666), beta-catenin-TCF complex (GO:1990907), G protein-coupled receptor binding (GO:0001664), and PDZ domain binding (GO:0030165) are among the top pathways based on adjusted p-value (Supplementary Tables 1-3). This information about hub genes and their involved enriched pathways can guide us in the way to understanding this signaling pathway better by *in vitro* analysis that ultimately can pave the way for more applied sciences in the clinical area.

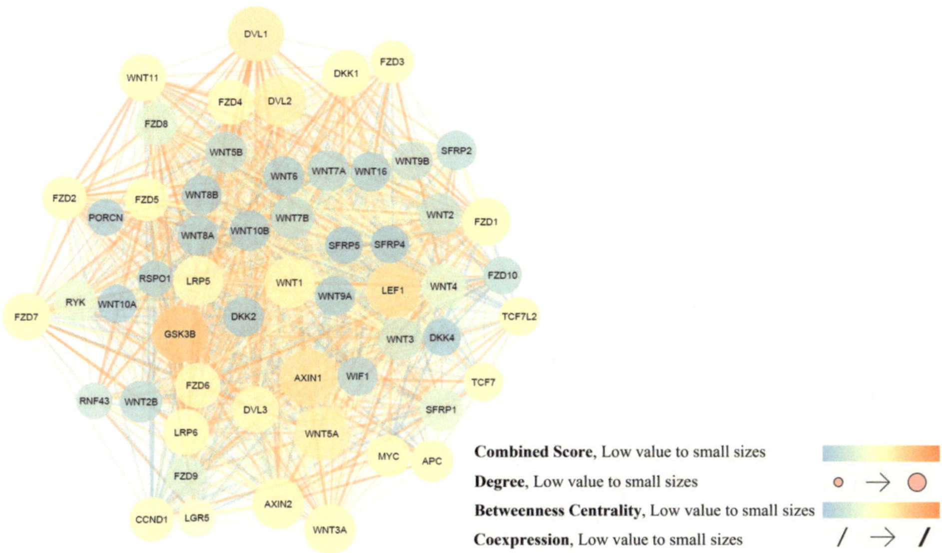

Figure 3.
PPI network of hub genes generated by bioinformatic analysis. Nodes and edges are visualized based on parameters weight such as Degree, Betweenness, Coexpression, and Combined score.

5. Clinical applications related to the Wnt signaling pathway

Improper activation of the Wnt signaling pathway is linked to many cancers, especially in the colon, liver, and breast. Conversely, reactivation of this pathway is essential for tissue regeneration following damage; insufficient activation results in regeneration failure and tissue degeneration. These phenomena underscore the critical roles of Wnt signaling in various tissues, highlighting that a thorough understanding of this pathway can inform numerous medical applications. Proper regulation of Wnt signaling is crucial for effective tissue regeneration, managing cancer progression, and preventing relapse, particularly concerning cancer stem cells. Most current anti-cancer drugs, including small molecule inhibitors and monoclonal antibodies (mAbs), are designed to target rapidly proliferating cancer cells. However, these therapies often fail to eliminate CSCs, which are critical for tumor maintenance, recurrence, and resistance to treatment. Given the failure of current therapies to eradicate CSCs, developing strategies to target CSCs selectively is crucial [135]. Potential approaches are targeting specific pathways in CSCs and microenvironment modulations. Many drugs have been developed for cancer treatment, each with a specific target. For example, LGK974 (Novartis) inhibits Porcupine, an enzyme essential for Wnt secretion. It is in Phase I clinical trials for various cancers, including melanoma and breast cancer [32, 136]. Compounds like XAV939 (Novartis) [137] and G007-LK (Roche) [138] prevent β-catenin accumulation by inhibiting TNK1 and TNK2, key in the β-catenin destruction complex. They show promise but may cause intestinal toxicity [138]. PRI-724 (Prism Pharma) blocks the CBP/β-catenin complex, inhibiting Wnt/β-catenin pathway genes, showing efficacy in AML and solid tumors [135]. Niclosamide, an anti-helminthic drug, inhibits Wnt/β-catenin signaling but requires improved bioavailability for oncology use [139].

Clofazimine, an anti-leprosy drug, is also a potent Wnt/β-catenin inhibitor [140]. OMP-54F28 (OncoMed) acts as a Wnt protein scavenger, reducing tumor growth without major side effects [141]. OTSA101, an antibody against FZD10, shows antitumor activity in synovial sarcomas [142]. Vantictumab (OncoMed) targets multiple FZD receptors, reducing tumor growth, but its clinical trials faced setbacks [12]. Antibodies targeting secreted frizzled-related protein 2 (sFRP2) show potential in reducing tumor volume and angiogenesis in cancers such as angiosarcoma and breast cancer [143]. The Wnt signaling pathway is essential for stem cell proliferation and plays a crucial role in regenerative processes across various tissues, including the liver, brain, muscle, skin, and bone. Consequently, there are ongoing efforts to harness the Wnt signaling pathway for clinical applications in tissue regeneration [135]. Research in this area has identified several molecules that can target the Wnt pathway for regenerative purposes. For example, the small molecule agonist 2-AMBMP has shown promising results in promoting liver regeneration in rats [135]. In bone regeneration, neutralizing antibodies against Wnt inhibitors like Dkk1 and sclerostin and the direct application of Wnt proteins or agonists have been effective [144]. Additionally, R-spondin2, Wnt10b, and liposomal Wnt3a have been found to enhance skin healing and stimulate hair growth [145–147]. In the field of neuroprotection, curcumin-coated nanoparticles as agonists of the canonical Wnt signaling pathway have demonstrated the potential to regulate neurogenesis [148]. These findings suggest that targeting the Wnt signaling pathway could lead to significant advancements in regenerative medicine and therapeutic interventions for various tissue injuries and diseases.

6. Wnt signaling pathway-based biomarkers and therapeutic drugs

Biomarkers in the Wnt signaling pathway play a crucial role in disease diagnosis, prognosis, and therapeutic response prediction. These biomarkers, including proteins, genetic mutations, and epigenetic modifications, help identify Wnt pathway activation levels, which are often linked to various cancers, neurodegenerative diseases, and developmental disorders. **Table 1** summarizes key biomarkers, their roles,

Biomarker	Function in Pathway	Disease Context	Reference
β-catenin (total)	Key effector stabilizes to enter the nucleus and activate transcription	High levels indicate active pathway, which is common in cancer	[149]
β-catenin (nuclear)	Transcriptional co-activator binds TCF/LEF to activate target genes	High levels indicate active pathway linked to cancer progression	[150]
Phospho-β-catenin	Marked for degradation by the destruction complex when the pathway is off	Low levels indicate active pathway, seen in cancers with pathway activation	[151]
APC mutations	Loss of function disrupts the destruction complex, leading to pathway activation	Predominant in colorectal cancer, also in other solid tumors	[152]
Axin mutations	Loss of function disrupts the destruction complex, activates the pathway	Found in some cancers, less common than APC mutations	[153]

Biomarker	Function in Pathway	Disease Context	Reference
GSK-3β activity	Phosphorylates β-catenin for degradation, part of the destruction complex	Low activity indicates an active pathway, relevant in cancer	[154]
Phospho-GSK-3β (inactivated)	Inactivated form reduces β-catenin degradation, promoting pathway activation	High levels indicate an active pathway, seen in regenerative contexts	[155]
Wnt ligands (e.g., Wnt1 and Wnt3a)	Secreted glycoproteins that bind frizzled receptors to activate the pathway	High expression indicates potential activation linked to cancer and development	[156]
Frizzled receptors (FZD1–10)	Receptors for Wnt ligands, initiate signaling	Expression levels correlate with pathway activity, relevant in cancer	[157]
LRP5	Co-receptor, enhance canonical pathway activation	Expression levels correlate with activity, implicated in bone diseases	[158]
DKK1	Secreted inhibitor, binds LRP5/6 to block pathway	Low levels may indicate an active pathway and high levels in some cancers	[159]
sFRP (soluble frizzled-related protein)	Binds Wnt ligands, inhibits the pathway	Low levels may indicate an active pathway relevant to cancer	[160]
WIF1 (Wnt inhibitory factor 1)	Binds Wnt ligands, inhibits the pathway	Low levels may indicate an active pathway, seen in cancer progression	[161]
TCF/LEF transcription factors	Bind β-catenin to activate target gene transcription	Expression levels are relevant and high in active pathway contexts	[162]
Wnt target genes (e.g., c-myc)	Downstream effectors, activated by β-catenin/TCF complex	High expression indicates an active pathway used in cancer diagnostics	[163]

Table 1.
Key biomarkers affecting Wnt signaling, their roles, and disease contexts.

Drug/Agent Name	Target	Mechanism	Reference
LGK974 (BBI-801)	Porcupine	Inhibits Wnt ligand secretion by blocking palmitoylation	[136]
ETC-159	Porcupine	Prevents Wnt ligand secretion	[164]
Vantictumab (OMP-18R5)	Frizzled receptors	Monoclonal antibody blocks Wnt binding to receptors	[165]
Foxy-5	Frizzled receptors	Synthetic peptide mimics Wnt ligands to activate the pathway	[166]
PRI-724	CBP/β-catenin interaction	Disrupts interaction blocks transcription of target genes	[167]
ICG-001	CBP/β-catenin interaction	CREB-binding protein (CBP) inhibitor disrupts β-catenin/TCF-mediated transcription	[168]
LF3	β-Ctenin/TCF interaction	β-Catenin/TCF4 interaction inhibitor	[169]

Drug/Agent Name	Target	Mechanism	Reference
Quercetin	β-Catenin/TCF interaction	β-Catenin/TCF4 interaction inhibitor	[170]
XAV939	Tankyrase	Stabilizes Axin by inhibiting its degradation, increases β-catenin degradation	[171]
G007-LK	Tankyrase	Stabilizes Axin, inhibits the pathway	[138]
JW55	Tankyrase	inhibits Tankyrase, promotes β-catenin degradation	[172]
Salinomycin	Wnt/β-catenin signaling	Target cancer stem cells inhibit the pathway	[173]
Niclosamide	Wnt/β-catenin signaling	Inhibits the pathway, repurposed from anti-parasitic use	[174]
Lithium	GSK-3β	Inhibits GSK-3β, prevents β-catenin phosphorylation, activates pathway	[175]
CHIR99021	GSK-3β	Inhibits GSK-3β, activates pathway	[176]
SB-216763	GSK-3β	Inhibits GSK-3β, activates pathway	[177]
DKN-01	DKK1	Monoclonal antibody targeting DKK1	[178]
BHQ880	DKK1	Monoclonal antibody targeting DKK1	[179]

Table 2.
Wnt signaling therapeutic drugs, their targets, and mechanisms.

and disease contexts. Understanding these biomarkers provides valuable insights into disease progression and aids in developing targeted therapies.

Therapeutic drugs targeting the Wnt pathway are designed to either inhibit or activate it, depending on the disease context. Inhibitors are primarily used in cancers where the pathway is overactive, while activators are explored for regenerative medicine and conditions like bone disorders. **Table 2** details key drugs, their targets, mechanisms, and indications. By modulating Wnt pathway activity, these drugs offer new avenues for precision medicine, with ongoing clinical trials evaluating their safety and efficacy across various conditions.

7. Future perspectives

The continuous advances in characterization of the Wnt signaling pathway give us a promising perspective of future research and medical uses. The advancing knowledge regarding the dynamics of Wnt signaling in various tissues. Such information will enable researchers to address the heterogeneity within stem cell populations and tumor microenvironments, thereby refining our understanding of cell fate decisions and the molecular underpinnings of disease progression.

In parallel, the development of novel molecular tools—such as CRISPR/Cas9-based gene editing and RNA interference strategies—offers the potential to modulate Wnt pathway components with high precision. This could lead to more effective strategies for tissue regeneration and targeted cancer therapies, especially in malignancies where dysregulated Wnt signaling has been screened. Future clinical trials are anticipated to evaluate the efficacy of these targeted interventions, particularly in contexts like regenerative medicine and cancer.

Another critical frontier lies in the comprehensive mapping of PPI networks and bioinformatics modeling techniques within the Wnt pathway. Advances in computational modeling and high-throughput proteomics are expected to identify novel regulatory nodes and feedback loops, potentially revealing new therapeutic targets. A deeper understanding of these networks may not only facilitate the design of combination therapies but also help overcome resistance mechanisms that limit the success of current treatments.

Finally, this interdisciplinary approach can lead to the translation of these molecular insights into clinical applications. Collaboration between biologists, chemists, bioinformaticians, and clinicians is essential to filling the gaps. By integrating systems biology with precision medicine, future research can pave the way for personalized medicine that leverages the full potential of modulating the Wnt signaling pathway.

The future of Wnt signaling research promises us breakthroughs that enhance our understanding of both normal developmental processes and pathological conditions, ultimately leading to innovative and more effective therapeutic strategies.

8. Conclusions

The Wnt signaling pathway is involved in many critical cellular activities. Recent research has identified the contribution of the Wnt signaling pathway in stem cell regulation, tissue homeostasis, and signaling transduction regulation. This knowledge about the Wnt signaling pathway guides us to know the underlying control mechanisms in different tissues and their cells. This knowledge guides us in identifying pathways and different elements of this signaling pathway that can be considered as a target of different molecules. However, the exact mechanisms controlling and affecting this signaling pathway have not been fully understood. The exact effect of Wnt ligands and their specific role in the activation of downstream signals still needs further research. Additionally, the crosstalk of the Wnt signaling pathway with other signaling pathways and its exact effect on the signaling cascade needs more investigation. Accomplishing more advances related to the signaling pathways, especially the Wnt signaling pathway, not only increases our understanding of regulation and development but also can make our search for anti-cancer drugs and regenerative medicine progress faster.

Acknowledgements

This research was sponsored by the Faculty of Biotechnology at Amol University of Special Modern Technologies in the Islamic Republic of Iran. This research received no external funding.

Conflict of interest

The authors declare no conflict of interest.

Author contributions

N.G: Assembly of data, designed the experiment, data analysis, wrote and edited the manuscript, and prepared **Figures 1–3**; H.A. Carried out and edited the chapter; T.S: Carried out and edited the chapter.

AI tools

Artificial intelligence tools are strictly prohibited from generating data, composing chapter sections, formulating scientific conclusions, or engaging in activities that could result in scientific misrepresentation. Their sole purpose is for editorial correction and grammar checks within the text, aimed at refining and enhancing clarity and accuracy. These tools are not permitted to manipulate data or misrepresent scientific discourse in any way.

Supplementary information

Additional supporting information may be found in the supporting information tab for this chapter: https://drive.google.com/drive/folders/1OHkTgiHP3oi-ZHBa__z60G8HMvI7zB7X?usp=drive_link

Abbreviations

BIO	6-bromoindirubin-3′-oxime
AP-1	activator protein-1
BCL9	B-cell lymphoma9
PIP2	biphosphates
BRG1	Brahma-related gene 1
CaN	calcineurin
CamKII	calcium/calmodulin-dependent kinase II
CAMKII	calcium/calmodulin-dependent protein kinase II
CSCs	cancer stem cells
CK1α	casein kinase 1α
C/EBP	CCAAT/enhancer binding protein
CBP	cyclic adenosine monophosphate response element (CREB)-binding protein
CRD	cysteine-rich domain
DAG	diacylglycerol
DVL	disheveled
ESCs	embryonic stem cells
ER	endoplasmic reticulum
ECM	extracellular matrix
ERK	extracellular signal-related kinase
FZD	Frizzled
GO	gene ontology
GSK-3β	glycogen synthase kinase 3 β
HFSCs	hair follicle stem cells

HSCs	hematopoietic stem cells
HSPGs	heparan sulfate proteoglycans
ICM	inner cell mass
IP3	inositol (1,4,5)-triphosphates
iPSCs	induced pluripotent stem cells
JNK	JUN-N-terminal kinase
LIF	leukemia inhibitory factor
MaSCs	mammary stem cells
MMTV	mammary tumor virus
MBOAT	membrane-bound O-acyltransferase
MSCs	mesenchymal stem cells
MAPK/ERK	mitogen-activated protein kinase/extracellular signal-related kinase
mAbs	monoclonal antibodies
MEF	mouse embryonic fibroblast feeder
NLK	nemo-like kinase
NSCs	neural stem cells
NFAT	nuclear factor of activated T-cells
NFAT	nuclear factor-activated T-cells
PPARγ	peroxisome proliferator-activated receptor γ
PLC	phospholipase C
PCP	planar cell polarity
PKC	protein kinase C
PTK7	protein tyrosine kinase 7
PPI networks	protein-protein interaction networks
ROR	RAR-related orphan receptor
RhoA	Ras homolog gene family member A
RAC	Ras-related C3 botulinum toxin substrate
Ryk	receptor-like tyrosine kinase
Ser	serine
SCF	Skp1-Cullin-F-box
TCF	T-cell factor
TCF/LEF	T-cell factor/lymphoid enhancer factor
TAK-1	TGF-β-Activated kinase 1
Thr41	threonine 41
TGN	Trans-Golgi network
Ube2c	ubiquitin-conjugating enzyme E2C
β-TrCP	β-transduction repeat-containing protein

Author details

Nima Ghasemi[1], Hossein Azizi[1]* and Thomas Skutella[2]

1 Faculty of Biotechnology, Amol University of Special Modern Technologies, Amol, Iran

2 Medical Faculty, Institute for Anatomy and Cell Biology, University of Heidelberg, Heidelberg, Germany

*Address all correspondence to: h.azizi@ausmt.ac.ir

References

[1] Ghasemi N, Azizi H, Skutella T. Unraveling the significance of Nanog in the generation of embryonic stem-like cells from spermatogonia stem cells: A combined in silico analysis and in vitro experimental approach. International Journal of Molecular Sciences. 2024;**25**. DOI: 10.3390/ijms25094833

[2] Jagannathan C, Waddington R, Nishio Ayre W. Nanoparticle and nanotopography-induced activation of the Wnt pathway in bone regeneration. Tissue Engineering Part B: Reviews. 2024;**30**(2):270-283

[3] Bhagtaney L, Dharmarajan A, Warrier S. miRNA on the battlefield of cancer: Significance in cancer stem cells, WNT pathway, and treatment. Cancers. 2024;**16**(5):957

[4] Clevers H, Loh KM, Nusse R. An integral program for tissue renewal and regeneration: Wnt signaling and stem cell control. Science. 2014;**346**(6205):1248012. DOI: 10.1126/science.1248012

[5] Nusse R, Varmus HE. Many tumors induced by the mouse mammary tumor virus contain a provirus integrated in the same region of the host genome. Cell. 1982;**31**(1):99-109. DOI: 10.1016/0092-8674(82)90409-3

[6] Rijsewijk F, Schuermann M, Wagenaar E, Parren P, Weigel D, Nusse R. The drosophila homolog of the mouse mammary oncogene int-1 is identical to the segment polarity gene wingless. Cell. 1987;**50**(4):649-657. DOI: 10.1016/0092-8674(87)90038-9

[7] Mohammed MK, Shao C, Wang J, Wei Q, Wang X, Collier Z, et al. Wnt/β-catenin signaling plays an ever-expanding role in stem cell self-renewal, tumorigenesis and cancer chemoresistance. Genes & Diseases. 2016;**3**(1):11-40. DOI: 10.1016/j.gendis.2015.12.004

[8] Tejeda-Muñoz N, Mei KC. Wnt signaling in cell adhesion, development, and colon cancer. IUBMB Life. 2024;**76**(7):383-396. DOI: 10.1002/iub.2806

[9] Katoh M, Katoh M. WNT signaling pathway and stem cell signaling network. Clinical Cancer Research. 2007;**13**(14):4042-4045. DOI: 10.1158/1078-0432.Ccr-06-2316

[10] Raslan AA, Yoon JK. WNT signaling in lung repair and regeneration. Molecules and Cells. 2020;**43**(9):774-783. DOI: 10.14348/molcells.2020.0059

[11] Zhang Y, Wang X. Targeting the Wnt/β-catenin signaling pathway in cancer. Journal of Hematology & Oncology. 2020;**13**(1):165

[12] Pećina-Šlaus N, Aničić S, Bukovac A, Kafka A. Wnt signaling inhibitors and their promising role in tumor treatment. International Journal of Molecular Sciences. 2023;**24**(7):6733. DOI: 10.3390/ijms24076733

[13] Zhang X, Li C, Wu Y, Cui P. The research progress of Wnt/β-catenin signaling pathway in colorectal cancer. Clinics and Research in Hepatology and Gastroenterology. 2023;**47**(3):102086

[14] Vlashi R, Zhang X, Wu M, Chen G. Wnt signaling: Essential roles in osteoblast differentiation, bone metabolism and therapeutic implications for bone and skeletal disorders. Genes & Diseases.

2023;**10**(4):1291-1317. DOI: 10.1016/j.gendis.2022.07.011

[15] Hao J, Han X, Huang H, Yu X, Fang J, Zhao J, et al. Sema3C signaling is an alternative activator of the canonical WNT pathway in glioblastoma. Nature Communications. 2023;**14**(1):2262. DOI: 10.1038/s41467-023-37397-w

[16] Clevers H, Nusse R. Wnt/β-catenin signaling and disease. Cell. 2012;**149**(6):1192-1205. DOI: 10.1016/j.cell.2012.05.012

[17] Cring M, Widhopf G, Ghia EM, Kipps TJ. Defining signaling interactomes involved in non-canonical Wnt signaling by ROR1 or ROR2 in hairy cell leukemia that can influence cell morphology and migration. Blood. 2023;**142**:4378. DOI: 10.1182/blood-2023-189562

[18] Walter NM, Ohki CMY, Rickli M., Smigielski L, Walitza S, Grünblatt E. An investigation on the alterations in Wnt signaling in ADHD across developmental stages. Neuroscience Applied. 2024;**3**:104070

[19] Miao C-G, Yang Y-Y, He X, Huang C, Huang Y, Zhang L, et al. Wnt signaling in liver fibrosis: Progress, challenges and potential directions. Biochimie. 2013;**95**(12):2326-2335

[20] Liu J, Xiao Q, Xiao J, Niu C, Li Y, Zhang X, et al. Wnt/β-catenin signalling: Function, biological mechanisms, and therapeutic opportunities. Signal Transduction and Targeted Therapy. 2022;**7**(1):3. DOI: 10.1038/s41392-021-00762-6

[21] Komekado H, Yamamoto H, Chiba T, Kikuchi A. Glycosylation and palmitoylation of Wnt-3a are coupled to produce an active form of Wnt-3a. Genes to Cells. 2007;**12**(4):521-534. DOI: 10.1111/j.1365-2443.2007.01068.x

[22] Willert K, Brown JD, Danenberg E, Duncan AW, Weissman IL, Reya T, et al. Wnt proteins are lipid-modified and can act as stem cell growth factors. Nature. 2003;**423**(6938):448-452. DOI: 10.1038/nature01611

[23] Yu J, Virshup DM. Functional regulation of Wnt protein through post-translational modifications. Biochemical Society Transactions. 2022;**50**(6):1797-1808. DOI: 10.1042/bst20220735

[24] Aponte Santamaria DF. The Role of Lipids in Wnt Secretion and Function. Diss. 2023. DOI: 10.11588/heidok.00031813

[25] Vecchia L. Wnt Secretion, Trafficking and Pathway Activation: Towards a Structure-Based Understanding and Modulation. University of Oxford; 2021

[26] Miura GI, Treisman JE. Lipid modification of secreted signaling proteins. Cell Cycle. 2006;**5**(11):1184-1188. DOI: 10.4161/cc.5.11.2804

[27] Smotrys JE, Linder ME. Palmitoylation of intracellular signaling proteins: Regulation and function. Annual Review of Biochemistry. 2004;**73**:559-587. DOI: 10.1146/annurev.biochem.73.011303.073954

[28] Gu Y, Liu X, Liao L, Gao Y, Shi Y, Ni J, et al. Relationship between lipid metabolism and hedgehog signaling pathway. The Journal of Steroid Biochemistry and Molecular Biology. 2021;**209**:105825

[29] Janda CY, Waghray D, Levin AM, Thomas C, Garcia KC. Structural basis

of Wnt recognition by frizzled. Science. 2012;**337**(6090):59-64. DOI: 10.1126/science.1222879

[30] Nile AH, Hannoush RN. Probing interaction of lipid-modified Wnt protein and its receptors by ELISA. Methods in Molecular Biology. 2019;**2009**:217-225. DOI: 10.1007/978-1-4939-9532-5_17

[31] Qin K, Yu M, Fan J, Wang H, Zhao P, Zhao G, et al. Canonical and noncanonical Wnt signaling: Multilayered mediators, signaling mechanisms and major signaling crosstalk. Genes & Diseases. 2024;**11**(1):103-134. DOI: 10.1016/j.gendis.2023.01.030

[32] Takada R, Satomi Y, Kurata T, Ueno N, Norioka S, Kondoh H, et al. Monounsaturated fatty acid modification of Wnt protein: Its role in Wnt secretion. Developmental Cell. 2006;**11**(6):791-801. DOI: 10.1016/j.devcel.2006.10.003

[33] Bagchi DP, MacDougald OA. Wnt signaling: From mesenchymal cell fate to lipogenesis and other mature adipocyte functions. Diabetes. 2021;**70**(7):1419-1430. DOI: 10.2337/dbi20-0015

[34] Mii Y, Takada S. Heparan sulfate proteoglycan clustering in Wnt signaling and dispersal. Frontiers in Cell and Developmental Biology. 2020;**8**:631

[35] Li B, Niswander LA. TMEM132A, a novel Wnt signaling pathway regulator through Wntless (WLS) interaction. Frontiers in Cell and Development Biology. 2020;**8**:599890. DOI: 10.3389/fcell.2020.599890

[36] Chai G, Szenker-Ravi E, Chung C, Li Z, Wang L, Khatoo M, et al. A human pleiotropic multiorgan condition caused by deficient Wnt secretion. The New England Journal of Medicine. 2021;**385**(14):1292-1301. DOI: 10.1056/NEJMoa2033911

[37] Lu D, Li Y, Liu Q-R, Wu Q, Zhang H, Xie P, et al. Wls promotes the proliferation of breast cancer cells via Wnt signaling. Medical Oncology. 2015;**32**:1-7

[38] Goodman RM, Thombre S, Firtina Z, Gray D, Betts D, Roebuck J, et al. Sprinter: A novel transmembrane protein required for Wg secretion and signaling. Development. 2006;**133**(24):4901-4911. DOI: 10.1242/dev.02674

[39] Xie S, Dierlam C, Smith E, Duran R, Williams A, Davis A, et al. The retromer complex regulates *C. elegans* development and mammalian ciliogenesis. Journal of Cell Science. 2022;**135**(10):jcs259396. DOI: 10.1242/jcs.259396

[40] Zhang P, Wu Y, Belenkaya TY, Lin X. SNX3 controls wingless/Wnt secretion through regulating retromer-dependent recycling of Wntless. Cell Research. 2011;**21**(12):1677-1690. DOI: 10.1038/cr.2011.167

[41] Franch-Marro X, Wendler F, Guidato S, Griffith J, Baena-Lopez A, Itasaki N, et al. Wingless secretion requires endosome-to-Golgi retrieval of Wntless/Evi/sprinter by the retromer complex. Nature Cell Biology. 2008;**10**(2):170-177. DOI: 10.1038/ncb1678

[42] Port F, Kuster M, Herr P, Furger E, Bänziger C, Hausmann G, et al. Wingless secretion promotes and requires retromer-dependent cycling of Wntless. Nature Cell Biology. 2008;**10**(2):178-185. DOI: 10.1038/ncb1687

[43] Brown HM, Murray SA, Northrup H, Au KS, Niswander LA. Snx3 is important

for mammalian neural tube closure via its role in canonical and non-canonical WNT signaling. Development. 2020;**147**(22):dev192518. DOI: 10.1242/dev.192518

[44] Nolo R, Abbott LA, Bellen HJ. Senseless, a Zn finger transcription factor, is necessary and sufficient for sensory organ development in drosophila. Cell. 2000;**102**(3):349-362. DOI: 10.1016/s0092-8674(00)00040-4

[45] Tabata T, Takei Y. Morphogens, their identification and regulation. Development. 2004;**131**(4):703-712. DOI: 10.1242/dev.01043

[46] van Amerongen R. Celebrating discoveries in Wnt signaling: How one man gave wings to an entire field. Cell. 2020;**181**(3):487-491. DOI: 10.1016/j.cell.2020.03.033

[47] Stapornwongkul KS, Vincent J-P. Generation of extracellular morphogen gradients: The case for diffusion. Nature Reviews Genetics. 2021;**22**(6):393-411. DOI: 10.1038/s41576-021-00342-y

[48] Rosenbauer J, Zhang C, Mattes B, Reinartz I, Wedgwood K, Schindler S, et al. Modeling of Wnt-mediated tissue patterning in vertebrate embryogenesis. PLoS Computational Biology. 2020;**16**(6):e1007417. DOI: 10.1371/journal.pcbi.1007417

[49] Mehta S, Hingole S, Chaudhary V. The emerging mechanisms of Wnt secretion and signaling in development. Frontiers in Cell and Development Biology. 2021;**9**:714746. DOI: 10.3389/fcell.2021.714746

[50] Tran THN, Takada R, Krayukhina E, Maruno T, Mii Y, Uchiyama S, et al. Soluble frizzled-related proteins promote exosome-mediated Wnt re-secretion. Communications Biology. 2024;**7**(1):254. DOI: 10.1038/s42003-024-05881-8

[51] Nie X, Wei X, Ma H, Fan L, Chen WD. The complex role of Wnt ligands in type 2 diabetes mellitus and related complications. Journal of Cellular and Molecular Medicine. 2021;**25**(14):6479-6495. DOI: 10.1111/jcmm.16663

[52] Gammons M, Bienz M. Multiprotein complexes governing Wnt signal transduction. Current Opinion in Cell Biology. 2018;**51**:42-49. DOI: 10.1016/j.ceb.2017.10.008

[53] Umar SA, Dong B, Nihal M, Chang H. Frizzled receptors in melanomagenesis: From molecular interactions to target identification. Frontiers in Oncology. 2022;**12**:1096134. DOI: 10.3389/fonc.2022.1096134

[54] Sharma M, Pruitt K. Wnt pathway: An integral hub for developmental and oncogenic signaling networks. International Journal of Molecular Sciences. 2020;**21**(21):8018

[55] Grainger S, Willert K. Mechanisms of Wnt signaling and control. WIREs Systems Biology and Medicine. 2018;**10**(5):e1422. DOI: 10.1002/wsbm.1422

[56] Ghosh N, Hossain U, Mandal A, Sil PC. The Wnt signaling pathway: A potential therapeutic target against cancer. Annals of the New York Academy of Sciences. 2019;**1443**(1):54-74. DOI: 10.1111/nyas.14027

[57] Houschyar KS, Tapking C, Borrelli MR, Popp D, Duscher D, Maan ZN, et al. Wnt pathway in bone repair and regeneration - What do we know so far. Frontiers in Cell and Development Biology. 2018;**6**:170. DOI: 10.3389/fcell.2018.00170

[58] Luo J, Yu Z, Tovar J, Nilsson A, Xu B. Critical review on anti-obesity effects of phytochemicals through Wnt/β-catenin signaling pathway. Pharmacological Research. 2022;**184**:106461. DOI: 10.1016/j.phrs.2022.106461

[59] Chambers M, Turki-Judeh W, Kim MW, Chen K, Gallaher SD, Courey AJ. Mechanisms of Groucho-mediated repression revealed by genome-wide analysis of Groucho binding and activity. BMC Genomics. 2017;**18**(1):215. DOI: 10.1186/s12864-017-3589-6

[60] Shah K, Kazi JU. Phosphorylation-dependent regulation of WNT/Beta-catenin signaling. Frontiers in Oncology. 2022;**12**:858782

[61] Zhu Y, X. Li advances of Wnt signalling pathway in colorectal cancer. Cells. 2023;**12**. DOI: 10.3390/cells12030447

[62] Lojk J, J. Marc roles of non-canonical Wnt signalling pathways in bone biology. International Journal of Molecular Sciences. 2021;**22**. DOI: 10.3390/ijms221910840

[63] Shi D-L. Wnt/planar cell polarity signaling controls morphogenetic movements of gastrulation and neural tube closure. Cellular and Molecular Life Sciences. 2022;**79**(12):586. DOI: 10.1007/s00018-022-04620-8

[64] Jansson L, Kim GS, Cheng AG. Making sense of Wnt signaling-linking hair cell regeneration to development. Frontiers in Cellular Neuroscience. 2015;**9**:66. DOI: 10.3389/fncel.2015.00066

[65] Martinez S, Scerbo P, Giordano M, Daulat AM, Lhoumeau AC, Thomé V, et al. The PTK7 and ROR2 protein receptors interact in the vertebrate WNT/planar cell polarity (PCP) pathway. The Journal of Biological Chemistry. 2015;**290**(51):30562-30572. DOI: 10.1074/jbc.M115.697615

[66] Najdi R, Holcombe RF, Waterman ML. Wnt signaling and colon carcinogenesis: Beyond APC. Journal of Carcinogenesis. 2011;**10**:5. DOI: 10.4103/1477-3163.78111

[67] Koni M, Pinnarò V, Brizzi MF. The Wnt signalling pathway: A tailored target in cancer. International Journal of Molecular Sciences. 2020;**21**(20). DOI: 10.3390/ijms21207697

[68] Hayat R, Manzoor M, Hussain A. Wnt signaling pathway: A comprehensive review. Cell Biology International. 2022;**46**(6):863-877. DOI: 10.1002/cbin.11797

[69] Park YG, Lee SE, Kim EY, Hyun H, Shin MY, Son YJ, et al. Effects of feeder cell types on culture of mouse embryonic stem cell in vitro. Development & Reproduction. 2015;**19**(3):119-126. DOI: 10.12717/dr.2015.19.3.119

[70] Hirai H, Karian P, Kikyo N. Regulation of embryonic stem cell self-renewal and pluripotency by leukaemia inhibitory factor. Biochemical Journal. 2011;**438**(1):11-23

[71] Boyer LA, Lee TI, Cole MF, Johnstone SE, Levine SS, Zucker JP, et al. Core transcriptional regulatory circuitry in human embryonic stem cells. Cell. 2005;**122**(6):947-956. DOI: 10.1016/j.cell.2005.08.020

[72] Okoye UC, Malbon CC, Wang HY. Wnt and frizzled RNA expression in human mesenchymal and embryonic (H7) stem cells. Journal of Molecular Signaling. 2008;**3**:16. DOI: 10.1186/1750-2187-3-16

[73] Li C, Furth EE, Rustgi AK, Klein PS. When you come to a fork in the road, take it: Wnt signaling activates multiple pathways through the APC/Axin/GSK-3 complex. Cells. 2023;**12**(18):2256

[74] Wen J, Liu J, Song G, Liu L, Tang B, Li Z. Effects of 6-bromoindirubin-3′-oxime on the maintenance of pluripotency of porcine embryonic germ cells in combination with stem cell factor, leukemia inhibitory factor and fibroblast growth factor. Reproduction. 2010;**139**(6):1039-1046. DOI: 10.1530/rep-09-0539

[75] Chen G, Guo YE, Li C, Li S, Wan X. Small molecules that promote self-renewal of stem cells and somatic cell reprogramming. Stem Cell Reviews and Reports. 2020;**16**(3):511-523. DOI: 10.1007/s12015-020-09965-w

[76] Hao J, Li T-G, Qi X, Zhao D-F, Zhao G-Q. WNT/β-catenin pathway up-regulates Stat3 and converges on LIF to prevent differentiation of mouse embryonic stem cells. Developmental Biology. 2006;**290**(1):81-91. DOI: 10.1016/j.ydbio.2005.11.011

[77] Ying QL, Wray J, Nichols J, Batlle-Morera L, Doble B, Woodgett J, et al. The ground state of embryonic stem cell self-renewal. Nature. 2008;**453**(7194):519-523. DOI: 10.1038/nature06968

[78] Nusse R, Fuerer C, Ching W, Harnish K, Logan C, Zeng A, et al. Wnt signaling and stem cell control. Cold Spring Harbor Symposia on Quantitative Biology. 2008;**73**:59-66. DOI: 10.1101/sqb.2008.73.035

[79] Pu X-Y, Zheng D-F, Lv T, Zhou Y-J, Yang J-Y, Jiang L. Overexpression of transcription factor 3 drives hepatocarcinoma development by enhancing cell proliferation via activating Wnt signaling pathway. Hepatobiliary & Pancreatic Diseases International. 2022;**21**(4):378-386

[80] Davidson KC, Adams AM, Goodson JM, McDonald CE, Potter JC, Berndt JD, et al. Wnt/β-catenin signaling promotes differentiation, not self-renewal, of human embryonic stem cells and is repressed by Oct4. Proceedings of the National Academy of Sciences of the United States of America. 2012;**109**(12):4485-4490. DOI: 10.1073/pnas.1118777109

[81] Blauwkamp TA, Nigam S, Ardehali R, Weissman IL, Nusse R. Endogenous Wnt signalling in human embryonic stem cells generates an equilibrium of distinct lineage-specified progenitors. Nature Communications. 2012;**3**:1070. DOI: 10.1038/ncomms2064

[82] Singh AM, Reynolds D, Cliff T, Ohtsuka S, Mattheyses AL, Sun Y, et al. Signaling network crosstalk in human pluripotent cells: A Smad2/3-regulated switch that controls the balance between self-renewal and differentiation. Cell Stem Cell. 2012;**10**(3):312-326. DOI: 10.1016/j.stem.2012.01.014

[83] Takam Kamga P, Bazzoni R, Dal Collo G, Cassaro A, Tanasi I, Russignan A, et al. The role of Notch and Wnt signaling in MSC communication in normal and leukemic bone marrow niche. Frontiers in Cell and Developmental Biology. 2020;**8**:599276. DOI: 10.3389/fcell.2020.599276

[84] de Winter TJJ, Nusse R. Running against the Wnt: How Wnt/β-catenin suppresses adipogenesis. Frontiers in Cell and Development Biology. 2021;**9**:627429. DOI: 10.3389/fcell.2021.627429

[85] Takada I, Suzawa M, Matsumoto K, Kato S. Suppression of PPAR transactivation switches cell fate of bone

marrow stem cells from adipocytes into osteoblasts. Annals of the New York Academy of Sciences. 2007;**1116**:182-195. DOI: 10.1196/annals.1402.034

[86] Jia Y, Liu Y, Liu S, Chang Y, Xu L, Li H. Wnt10b knockdown promotes UCP1 expression in brown adipose tissue in mice. Genes to Cells. 2023;**28**(11):764-775. DOI: 10.1111/gtc.13064

[87] Liu Y, Fang J, Zhang Q, Zhang X, Cao Y, Chen W, et al. Wnt10b-overexpressing umbilical cord mesenchymal stem cells promote critical size rat calvarial defect healing by enhanced osteogenesis and VEGF-mediated angiogenesis. Journal of Orthopaedic Translation. 2020;**23**:29-37. DOI: 10.1016/j.jot.2020.02.009

[88] Stürznickel J, Rolvien T, Delsmann A, Butscheidt S, Barvencik F, Mundlos S, et al. Clinical phenotype and relevance of LRP5 and LRP6 variants in patients with early-onset osteoporosis (EOOP). Journal of Bone and Mineral Research. 2021;**36**(2):271-282. DOI: 10.1002/jbmr.4197

[89] Ren, Na et al. Clinical features, treatment, and follow-up of OPPG and high-bone-mass disorders: LRP5 is a key regulator of bone mass. Osteoporosis international: A journal established as result of cooperation between the European Foundation for Osteoporosis and the National Osteoporosis Foundation of the USA. 2024;**35**(8):1395-1406. DOI: 10.1007/s00198-024-07080-x

[90] Kamizaki K, Endo M, Minami Y, Kobayashi Y. Role of noncanonical Wnt ligands and Ror-family receptor tyrosine kinases in the development, regeneration, and diseases of the musculoskeletal system. Developmental Dynamics. 2021;**250**(1):27-38. DOI: 10.1002/dvdy.151

[91] Van Camp JK, Beckers S, Zegers D, Van Hul W. Wnt signaling and the control of human stem cell fate. Stem Cell Reviews and Reports. 2014;**10**(2):207-229. DOI: 10.1007/s12015-013-9486-8

[92] Liu G, Vijayakumar S, Grumolato L, Arroyave R, Qiao H, Akiri G, et al. Canonical Wnts function as potent regulators of osteogenesis by human mesenchymal stem cells. Journal of Cell Biology. 2009;**185**(1):67-75. DOI: 10.1083/jcb.200810137

[93] Song H, Park K-H. Regulation and function of SOX9 during cartilage development and regeneration. Seminars in Cancer Biology. 2020;**67**:12-23. DOI: 10.1016/j.semcancer.2020.04.008

[94] Schmidt M, Poser C, Janster C, von Maltzahn J. The hairpin region of WNT7A is sufficient for binding to the Frizzled7 receptor and to elicit signaling in myogenic cells. Computational and Structural Biotechnology Journal. 2022;**20**:6348-6359. DOI: 10.1016/j.csbj.2022.10.047

[95] Huang YL, Anvarian Z, Döderlein G, Acebron SP, Niehrs C. Maternal Wnt/STOP signaling promotes cell division during early Xenopus embryogenesis. Proceedings of the National Academy of Sciences of the United States of America. 2015;**112**(18):5732-5737. DOI: 10.1073/pnas.1423533112

[96] Niehrs C, Da Silva F, Seidl C. Cilia as Wnt signaling organelles. Trends Cell Biol. 2025;**35**(1):24-32. DOI: 10.1016/j.tcb.2024.04.001

[97] Zi Z, Zhang Z, Li Q, An W, Zeng L, Gao D, et al. CCNYL1, but not CCNY, cooperates with CDK16 to regulate spermatogenesis in mouse. PLoS Genetics. 2015;**11**(8):e1005485. DOI: 10.1371/journal.pgen.1005485

[98] Acebron SP, Niehrs C. B-Catenin-independent roles of Wnt/LRP6 signaling. Trends in Cell Biology. 2016;**26**(12):956-967. DOI: 10.1016/j.tcb.2016.07.009

[99] Richter J, Traver D, Willert K. The role of Wnt signaling in hematopoietic stem cell development. Critical Reviews in Biochemistry and Molecular Biology. 2017;**52**(4):414-424. DOI: 10.1080/10409238.2017.1325828

[100] Nemeth MJ, Bodine DM. Regulation of hematopoiesis and the hematopoietic stem cell niche by Wnt signaling pathways. Cell Research. 2007;**17**(9):746-758. DOI: 10.1038/cr.2007.69

[101] Luis TC, Naber BA, Roozen PP, Brugman MH, de Haas EF, Ghazvini M, et al. Canonical wnt signaling regulates hematopoiesis in a dosage-dependent fashion. Cell Stem Cell. 2011;**9**(4):345-356. DOI: 10.1016/j.stem.2011.07.017

[102] Heinonen KM, Vanegas JR, Lew D, Krosl J, Perreault C. Wnt4 enhances murine hematopoietic progenitor cell expansion through a planar cell polarity-like pathway. PLoS One. 2011;**6**(4):e19279. DOI: 10.1371/journal.pone.0019279

[103] Reya T, Duncan AW, Ailles L, Domen J, Scherer DC, Willert K, et al. A role for Wnt signalling in self-renewal of haematopoietic stem cells. Nature. 2003;**423**(6938):409-414. DOI: 10.1038/nature01593

[104] Bhavanasi D, Klein PS. Wnt signaling in normal and malignant stem cells. Current Stem Cell Reports. 2016;**2**(4):379-387. DOI: 10.1007/s40778-016-0068-y

[105] Mulligan KA, Cheyette BN. Wnt signaling in vertebrate neural development and function. Journal of Neuroimmune Pharmacology. 2012;**7**(4):774-787. DOI: 10.1007/s11481-012-9404-x

[106] Abreu de Oliveira WA, El Laithy Y, Bruna A, Annibali D, Lluis F. Wnt signaling in the breast: From development to disease. Frontiers in Cell and Developmental Biology. 2022;**10**:884467

[107] Zeng YA, Nusse R. Wnt proteins are self-renewal factors for mammary stem cells and promote their long-term expansion in culture. Cell Stem Cell. 2010;**6**(6):568-577. DOI: 10.1016/j.stem.2010.03.020

[108] Yu Z, Pestell TG, Lisanti MP, Pestell RG. Cancer stem cells. The International Journal of Biochemistry & Cell Biology. 2012;**44**(12):2144-2151. DOI: 10.1016/j.biocel.2012.08.022

[109] Tai D, Wells K, Arcaroli J, Vanderbilt C, Aisner DL, Messersmith WA, et al. Targeting the WNT signaling pathway in cancer therapeutics. The Oncologist. 2015;**20**(10):1189-1198. DOI: 10.1634/theoncologist.2015-0057

[110] Khalaf AM, Fuentes D, Morshid AI, Burke MR, Kaseb AO, Hassan M, et al. Role of Wnt/β-catenin signaling in hepatocellular carcinoma, pathogenesis, and clinical significance. Journal of Hepatocellular Carcinoma. 2018;**5**:61-73. DOI: 10.2147/JHC.S156701

[111] Yu C, Yao Z, Jiang Y, Keller ET. Prostate cancer stem cell biology. Minerva Urologica e Nefrologica. 2012;**64**(1):19-33

[112] Liu S, U KP, Zhang J, et al. R-spodin2 enhances canonical Wnt signaling to maintain the stemness

of glioblastoma cells. Cancer Cell International. 2018;**18**:156. DOI: 10.1186/s12935-018-0655-3

[113] Sarabia-Sánchez MA, Moreno-Londoño AP, Castañeda-Patlán MC, Alvarado-Ortiz E, Martínez-Morales JC, Robles-Flores M. Non-canonical Wnt/Ca2+ signaling is essential to promote self-renewal and proliferation in colon cancer stem cells. Frontiers in Oncology. 2023;**13**:1121787. DOI: 10.3389/fonc.2023.1121787

[114] Yang J, Zhang K, Wu J, Shi J, Xue J, Li J, et al. Wnt5a increases properties of lung cancer stem cells and resistance to cisplatin through activation of Wnt5a/PKC signaling pathway. Stem Cells International. 2016;**2016**:1690896. DOI: 10.1155/2016/1690896

[115] Yin P, Bai Y, Wang Z, Sun Y, Gao J, Na L, et al. Non-canonical Fzd7 signaling contributes to breast cancer mesenchymal-like stemness involving Col6a1. Cell Communication and Signaling: CCS. 2020;**18**(1):143. DOI: 10.1186/s12964-020-00646-2

[116] Cantilena S, Pastorino F, Pezzolo A, Chayka O, Pistoia V, Ponzoni M, et al. Frizzled receptor 6 marks rare, highly tumourigenic stem-like cells in mouse and human neuroblastomas. Oncotarget. 2011;**2**(12):976-983. DOI: 10.18632/oncotarget.410

[117] Tan DWM, Barker N. Intestinal stem cells and their defining niche. Current topics in developmental biology. 2014;**107**:77-107

[118] Gehart H, Clevers H. Tales from the crypt: New insights into intestinal stem cells. Nature Reviews. Gastroenterology & Hepatology. 2019;**16**(1):19-34. DOI: 10.1038/s41575-018-0081-y

[119] Flanagan DJ, Austin CR, Vincan E, Phesse TJ. Wnt signalling in gastrointestinal epithelial stem cells. Genes (Basel). 2018;**9**(4). DOI: 10.3390/genes9040178

[120] Valenta T, Degirmenci B, Moor AE, Herr P, Zimmerli D, Moor MB, et al. Wnt ligands secreted by subepithelial mesenchymal cells are essential for the survival of intestinal stem cells and gut homeostasis. Cell Reports. 2016;**15**(5):911-918. DOI: 10.1016/j.celrep.2016.03.088

[121] Swoboda J, Mittelsdorf P, Chen Y, Weiskirchen R, Stallhofer J, Schüle S, et al. Intestinal Wnt in the transition from physiology to oncology. World Journal of Clinical Oncology. 2022;**13**(3):168

[122] Kosinski C, Li VS, Chan AS, Zhang J, Ho C, Tsui WY, et al. Gene expression patterns of human colon tops and basal crypts and BMP antagonists as intestinal stem cell niche factors. Proceedings of the National Academy of Sciences of the United States of America. 2007;**104**(39):15418-15423. DOI: 10.1073/pnas.0707210104

[123] Mei X, Gu M, Li M. Plasticity of Paneth cells and their ability to regulate intestinal stem cells. Stem Cell Research & Therapy. 2020;**11**(1):349. DOI: 10.1186/s13287-020-01857-7

[124] Sailaja BS, He XC, Li L. The regulatory niche of intestinal stem cells. The Journal of Physiology. 2016;**594**(17):4827-4836. DOI: 10.1113/jp271931

[125] Farin HF, Jordens I, Mosa MH, Basak O, Korving J, Tauriello DV, et al. Visualization of a short-range Wnt gradient in the intestinal stem-cell niche. Nature. 2016;**530**(7590):340-343. DOI: 10.1038/nature16937

[126] Yu M, Qin K, Fan J, Zhao G, Zhao P, Zeng W, et al. The evolving roles of

Wnt signaling in stem cell proliferation and differentiation, the development of human diseases, and therapeutic opportunities. Genes & Diseases. 2024;**11**(3):101026. DOI: 10.1016/j.gendis.2023.04.042

[127] Hawkshaw NJ, Hardman JA, Alam M, Jimenez F, Paus R. Deciphering the molecular morphology of the human hair cycle: Wnt signalling during the telogen-anagen transformation. The British Journal of Dermatology. 2020;**182**(5):1184-1193. DOI: 10.1111/bjd.18356

[128] Genander M, Cook PJ, Ramsköld D, Keyes BE, Mertz AF, Sandberg R, et al. BMP signaling and its pSMAD1/5 target genes differentially regulate hair follicle stem cell lineages. Cell Stem Cell. 2014;**15**(5):619-633. DOI: 10.1016/j.stem.2014.09.009

[129] Polkoff KM, Gupta NK, Green AJ, Murphy Y, Chung J, Gleason KL, et al. LGR5 is a conserved marker of hair follicle stem cells in multiple species and is present early and throughout follicle morphogenesis. Scientific Reports. 2022;**12**(1):9104. DOI: 10.1038/s41598-022-13056-w

[130] Jaks V, Barker N, Kasper M, van Es JH, Snippert HJ, Clevers H, et al. Lgr5 marks cycling, yet long-lived, hair follicle stem cells. Nature Genetics. 2008;**40**(11):1291-1299. DOI: 10.1038/ng.239

[131] Barker N, Huch M, Kujala P, van de Wetering M, Snippert HJ, van Es JH, et al. Lgr5(+ve) stem cells drive self-renewal in the stomach and build long-lived gastric units in vitro. Cell Stem Cell. 2010;**6**(1):25-36. DOI: 10.1016/j.stem.2009.11.013

[132] McDonald SA, Greaves LC, Gutierrez-Gonzalez L, Rodriguez-Justo M, Deheragoda M, Leedham SJ, et al. Mechanisms of field cancerization in the human stomach: The expansion and spread of mutated gastric stem cells. Gastroenterology. 2008;**134**(2):500-510. DOI: 10.1053/j.gastro.2007.11.035

[133] Kanehisa M, Goto S. KEGG: Kyoto encyclopedia of genes and genomes. Nucleic Acids Research. 2000;**28**(1):27-30. DOI: 10.1093/nar/28.1.27

[134] Kanehisa M, Furumichi M, Sato Y, Matsuura Y, Ishiguro-Watanabe M. KEGG: Biological systems database as a model of the real world. Nucleic Acids Research. 2025;**53**(D1):D672-d677. DOI: 10.1093/nar/gkae909

[135] Blagodatski A, Poteryaev D, Katanaev VL. Targeting the Wnt pathways for therapies. Molecular and Cellular Therapies. 2014;**2**:28. DOI: 10.1186/2052-8426-2-28

[136] Liu J, Pan S, Hsieh MH, Ng N, Sun F, Wang T, et al. Targeting Wnt-driven cancer through the inhibition of porcupine by LGK974. Proceedings of the National Academy of Sciences of the United States of America. 2013;**110**(50):20224-20229. DOI: 10.1073/pnas.1314239110

[137] Huang SM, Mishina YM, Liu S, et al. Tankyrase inhibition stabilizes axin and antagonizes Wnt signalling. Nature. 2009;**461**(7264):614-620. DOI: 10.1038/nature08356

[138] Lau T, Chan E, Callow M, Waaler J, Boggs J, Blake RA, et al. A novel tankyrase small-molecule inhibitor suppresses APC mutation-driven colorectal tumor growth. Cancer Research. 2013;**73**(10):3132-3144. DOI: 10.1158/0008-5472.Can-12-4562

[139] Jiang H, Li AM, Ye J. The magic bullet: Niclosamide. Frontiers

in oncology. 2022;**12**:1004978. DOI: 10.3389/fonc.2022.1004978

[140] Xu J, Koval A, Katanaev VL. Clofazimine: A journey of a drug. Biomedicine & Pharmacotherapy. 2023;**167**:115539. DOI: 10.1016/j.biopha.2023.115539

[141] Selvaggi F, Catalano T, Cotellese R, G.M. Aceto targeting Wnt/β-catenin pathways in primary liver tumours: From microenvironment signaling to therapeutic agents. Cancers. 2022;**14**. DOI: 10.3390/cancers14081912

[142] Giraudet A-L, Cassier PA, Iwao-Fukukawa C, Garin G, Badel J-N, Kryza D, et al. A first-in-human study investigating biodistribution, safety and recommended dose of a new radiolabeled MAb targeting FZD10 in metastatic synovial sarcoma patients. BMC Cancer. 2018;**18**(1):646. DOI: 10.1186/s12885-018-4544-x

[143] van Loon K, Huijbers EJM, Griffioen AW. Secreted frizzled-related protein 2: A key player in noncanonical Wnt signaling and tumor angiogenesis. Cancer and Metastasis Reviews. 2021;**40**(1):191-203. DOI: 10.1007/s10555-020-09941-3

[144] Hoeppner LH, Secreto FJ, Westendorf JJ. Wnt signaling as a therapeutic target for bone diseases. Expert Opinion on Therapeutic Targets. 2009;**13**(4):485-496. DOI: 10.1517/14728220902841961

[145] Chua AW, Ma D, Gan SU, Fu Z, Han HC, Song C, et al. The role of R-spondin2 in keratinocyte proliferation and epidermal thickening in keloid scarring. Journal of Investigative Dermatology. 2011;**131**(3):644-654

[146] Lei M, Guo H, Qiu W, Lai X, Yang T, Widelitz RB, et al. Modulating hair follicle size with W nt10b/DKK 1 during hair regeneration. Experimental Dermatology. 2014;**23**(6):407-413

[147] Whyte JL, Smith AA, Liu B, Manzano WR, Evans ND, Dhamdhere GR, et al. Augmenting endogenous Wnt signaling improves skin wound healing. PLoS One. 2013;**8**(10):e76883

[148] Liczbiński P, Michałowicz J, Bukowska B. Molecular mechanism of curcumin action in signaling pathways: Review of the latest research. Phytotherapy Research. 2020;**34**(8):1992-2005

[149] Shang S, Hua F, Hu ZW. The regulation of β-catenin activity and function in cancer: Therapeutic opportunities. Oncotarget. 2017;**8**(20):33972-33989. DOI: 10.18632/oncotarget.15687

[150] Wong SC, Lo ES, Chan AK, Lee KC, Hsiao WL. Nuclear beta catenin as a potential prognostic and diagnostic marker in patients with colorectal cancer from Hong Kong. Molecular Pathology. 2003;**56**(6):347-352. DOI: 10.1136/mp.56.6.347

[151] Uzun S, Isik A, Katipoglu K, Guner G, Akyol A. Characterization of the subcellular distribution of phospho-β-catenin in colorectal cancer. In Vivo. 2023;**37**(4):1576-1583. DOI: 10.21873/invivo.13242

[152] Kwong LN, Dove WF. APC and its modifiers in colon cancer. Advances in Experimental Medicine and Biology. 2009;**656**:85-106. DOI: 10.1007/978-1-4419-1145-2_8

[153] Salahshor S, Woodgett JR. The links between axin and carcinogenesis. Journal of Clinical Pathology. 2005;**58**(3):225-236. DOI: 10.1136/jcp.2003.009506

[154] Domoto T, Uehara M, Bolidong D, Minamoto T. Glycogen synthase kinase 3β in cancer biology and treatment. Cells. 2020;**9**(6). DOI: 10.3390/cells9061388

[155] Domoto T, Pyko IV, Furuta T, Miyashita K, Uehara M, Shimasaki T, et al. Glycogen synthase kinase-3β is a pivotal mediator of cancer invasion and resistance to therapy. Cancer Science. 2016;**107**(10):1363-1372. DOI: 10.1111/cas.13028

[156] Werner J, Boonekamp KE, Zhan T, Boutros M. The roles of secreted Wnt ligands in cancer. International Journal of Molecular Sciences. 2023;**24**(6). DOI: 10.3390/ijms24065349

[157] Ueno K, Hirata H, Hinoda Y, Dahiya R. Frizzled homolog proteins, microRNAs and Wnt signaling in cancer. International Journal of Cancer. 2013;**132**(8):1731-1740. DOI: 10.1002/ijc.27746

[158] Joiner DM, Ke J, Zhong Z, Xu HE, Williams BO. LRP5 and LRP6 in development and disease. Trends in Endocrinology and Metabolism. 2013;**24**(1):31-39. DOI: 10.1016/j.tem.2012.10.003

[159] Niehrs C. Function and biological roles of the Dickkopf family of Wnt modulators. Oncogene. 2006;**25**(57):7469-7481. DOI: 10.1038/sj.onc.1210054

[160] Bovolenta P, Esteve P, Ruiz JM, Cisneros E, Lopez-Rios J. Beyond Wnt inhibition: New functions of secreted frizzled-related proteins in development and disease. Journal of Cell Science. 2008;**121**(Pt 6):737-746. DOI: 10.1242/jcs.026096

[161] Wissmann C, Wild PJ, Kaiser S, Roepcke S, Stoehr R, Woenckhaus M, et al. WIF1, a component of the Wnt pathway, is down-regulated in prostate, breast, lung, and bladder cancer. The Journal of Pathology. 2003;**201**(2):204-212. DOI: 10.1002/path.1449

[162] Koelman EMR, Yeste-Vázquez A, Grossmann TN. Targeting the interaction of β-catenin and TCF/LEF transcription factors to inhibit oncogenic Wnt signaling. Bioorganic & Medicinal Chemistry. 2022;**70**:116920. DOI: 10.1016/j.bmc.2022.116920

[163] He T-C, Sparks AB, Rago C, Hermeking H, Zawel L, da Costa LT, et al. Identification of c-*MYC* as a target of the APC pathway. Science. 1998;**281**(5382):1509-1512. DOI: 10.1126/science.281.5382.1509

[164] Shah K, Panchal S, Patel B. Porcupine inhibitors: Novel and emerging anti-cancer therapeutics targeting the Wnt signaling pathway. Pharmacological Research. 2021;**167**:105532. DOI: 10.1016/j.phrs.2021.105532

[165] Gurney A, Axelrod F, Bond CJ, Cain J, Chartier C, Donigan L, et al. Wnt pathway inhibition via the targeting of frizzled receptors results in decreased growth and tumorigenicity of human tumors. Proceedings of the National Academy of Sciences of the United States of America. 2012;**109**(29):11717-11722. DOI: 10.1073/pnas.1120068109

[166] Säfholm A, Tuomela J, Rosenkvist J, Dejmek J, Härkönen P, Andersson T. The Wnt-5a–derived hexapeptide Foxy-5 inhibits breast cancer metastasis In vivo by targeting cell motility. Clinical

Cancer Research. 2008;**14**(20):6556-6563. DOI: 10.1158/1078-0432.CCR-08-0711

[167] Schmidtova S, Kalavska K, Liskova V, Plava J, Miklikova S, Kucerova L, et al. Targeting of deregulated Wnt/β-catenin Signaling by PRI-724 and LGK974 inhibitors in germ cell tumor cell lines. International Journal of Molecular Sciences. 2021;**22**(8). DOI: 10.3390/ijms22084263

[168] Gang EJ, Hsieh YT, Pham J, Zhao Y, Nguyen C, Huantes S, et al. Small-molecule inhibition of CBP/catenin interactions eliminates drug-resistant clones in acute lymphoblastic leukemia. Oncogene. 2014;**33**(17):2169-2178. DOI: 10.1038/onc.2013.169

[169] Fang L, Zhu Q, Neuenschwander M, Specker E, Wulf-Goldenberg A, Weis WI, et al. A small-molecule antagonist of the β-catenin/TCF4 interaction blocks the self-renewal of cancer stem cells and suppresses tumorigenesis. Cancer Research. 2016;**76**(4):891-901. DOI: 10.1158/0008-5472.Can-15-1519

[170] Park CH, Chang JY, Hahm ER, Park S, Kim HK, Yang CH. Quercetin, a potent inhibitor against beta-catenin/Tcf signaling in SW480 colon cancer cells. Biochemical and Biophysical Research Communications. 2005;**328**(1):227-234. DOI: 10.1016/j.bbrc.2004.12.151

[171] Huang S-MA, Mishina YM, Liu S, Cheung A, Stegmeier F, Michaud GA, et al. Tankyrase inhibition stabilizes axin and antagonizes Wnt signalling. Nature. 2009;**461**(7264):614-620. DOI: 10.1038/nature08356

[172] Waaler J, Machon O, Tumova L, Dinh H, Korinek V, Wilson SR, et al. A novel tankyrase inhibitor decreases canonical Wnt signaling in colon carcinoma cells and reduces tumor growth in conditional APC mutant mice. Cancer Research. 2012;**72**(11):2822-2832. DOI: 10.1158/0008-5472.Can-11-3336

[173] Lu D, Choi MY, Yu J, Castro JE, Kipps TJ, Carson DA. Salinomycin inhibits Wnt signaling and selectively induces apoptosis in chronic lymphocytic leukemia cells. Proceedings of the National Academy of Sciences. 2011;**108**(32):13253-13257. DOI: 10.1073/pnas.1110431108

[174] Pan JX, Ding K, Wang CY. Niclosamide, an old antihelminthic agent, demonstrates antitumor activity by blocking multiple signaling pathways of cancer stem cells. Chinese Journal of Cancer. 2012;**31**(4):178-184. DOI: 10.5732/cjc.011.10290

[175] Snitow ME, Bhansali RS, Klein PS. Lithium and therapeutic targeting of GSK-3. Cells. 2021;**10**(2). DOI: 10.3390/cells10020255

[176] Wang B, Khan S, Wang P, Wang X, Liu Y, Chen J, et al. A highly selective GSK-3β inhibitor CHIR99021 promotes Osteogenesis by activating canonical and autophagy-mediated Wnt signaling. Frontiers in Endocrinology. 2022;**13**:926622. DOI: 10.3389/fendo.2022.926622

[177] Coghlan MP, Culbert AA, Cross DA, Corcoran SL, Yates JW, Pearce NJ, et al. Selective small molecule inhibitors of glycogen synthase kinase-3 modulate glycogen metabolism and gene transcription. Chemistry & Biology. 2000;**7**(10):793-803. DOI: 10.1016/s1074-5521(00)00025-9

[178] Arend R, Dholakia J, Castro C, Matulonis U, Hamilton E, Jackson CG,

et al. DKK1 is a predictive biomarker for response to DKN-01: Results of a phase 2 basket study in women with recurrent endometrial carcinoma. Gynecologic Oncology. 2023;**172**:82-91. DOI: 10.1016/j.ygyno.2023.03.013

[179] Goldstein SD, Trucco M, Bautista Guzman W, Hayashi M, Loeb DM. A monoclonal antibody against the Wnt signaling inhibitor dickkopf-1 inhibits osteosarcoma metastasis in a preclinical model. Oncotarget. 2016;**7**(16):21114-21123. DOI: 10.18632/oncotarget.8522

Chapter 3

Cardiovascular Disease Genetic Risk Prediction Models: A Systematic Review

Sumera Khan and Xianquan Zhan

Abstract

This review provides a comprehensive evaluation of genetic cardiovascular disease (CVD) risk prediction models developed or validated for the general population. The study aims to assess these models' methodological quality, predictive performance, and validation approaches. A systematic search of electronic databases was conducted to identify studies published between 1990 and 2024. Data on study design, participant characteristics, predictor variables, statistical modeling approaches, validation methods, and performance metrics were extracted and analyzed. A total of nine studies met the inclusion criteria. Among them, five studies (56%) focused on model development, two studies (22%) performed internal validation, and two studies (22%) were externally validated. The most frequently included predictors were age (n = 9; 100%), sex (n = 7; 78%), LDL cholesterol (n = 6; 67%), and diabetes (n = 5; 56%). The genetic variants most commonly integrated into the models included APOE (n = 4), LPA (n = 5), and PCSK9 (n = 4). The AUROC scores ranged from 0.72 to 0.83, with Hosmer–Lemeshow calibration reported in five studies (56%). Cross-validation was the predominant validation method utilized in 56% of studies, whereas external validation was applied in only 22%. Although polygenic risk scores (PRS) have improved cardiovascular risk prediction, methodological inconsistencies, limited external validation, and lack of standardization remain key challenges. Future research should prioritize large-scale external validation studies, incorporate broader population-based datasets, and establish standardized performance evaluation frameworks to enhance model reliability and clinical applicability.

Keywords: cardiovascular disease, risk prediction, genetic risk, polygenic risk score, model performance evaluation

1. Introduction

Cardiovascular diseases (CVD) remain the leading cause of death globally. According to the World Health Organization (WHO), CVDs are responsible for approximately 17.9 million deaths each year, accounting for 32% of all global deaths [1]. Of these deaths, 85% are due to heart attacks and strokes. In 2021, ischemic heart disease alone caused 9 million deaths worldwide, making it the single largest cause of death [2]. Identifying individuals at high risk enables timely interventions, which can prevent serious illness and guide therapeutic decision-making. The global burden of CVD is expected to rise further, especially in low- and middle-income countries (LMICs), where healthcare infrastructure is often under-resourced and early detection tools are limited [3]. This trend is compounded by increasing rates of obesity, diabetes, sedentary lifestyles, and aging populations.

Over the last two decades, a few of the risk prediction models have been developed for predicting CVDs. The most commonly known, well-validated, and widely used conventional risk prediction models in clinical settings are Framingham Risk Score (FRS) [4], Reynolds Risk Score (RRS) [5, 6], Systematic Coronary Risk Evaluation (SCORE) [7], Prospective Cardiovascular Münster (PROCAM) Score [8], American College of Cardiology/American Heart Association 2013 risk score (ACC/AHA13) [9], and QRESEARCH cardiovascular risk (QRISK1 and QRISK2) [10]. Additionally, the Joint British Societies (JBS3) Risk Calculator [11] provides a lifetime risk assessment, offering a broader perspective on cardiovascular disease prevention. These models are often used in clinical practice to identify high-risk individuals and guide preventive interventions such as statin therapy or lifestyle modification [12]. However, these tools are typically based on a limited number of clinical and lifestyle factors and may fail to detect risk in younger individuals or those without obvious risk markers [13]. These traditional models help individuals and clinicians by identifying those at an increased risk of future cardiovascular events, allowing for timely lifestyle modifications and preventive medical treatment. However, while these models rely on clinical and lifestyle factors, they do not account for an individual's genetic predisposition to cardiovascular diseases (CVD). This omission is increasingly seen as a limitation, as studies have demonstrated that genetic risk can influence cardiovascular outcomes independently of conventional risk factors. Moreover, some individuals with low traditional risk scores still go on to experience cardiovascular events, highlighting the need for more precise predictive tools [14].

Genetic risk scores (GRS), particularly polygenic risk scores (PRS), have emerged as powerful tools for assessing an individual's genetic predisposition to cardiovascular disease (CVD). Unlike conventional risk models that rely on clinical and lifestyle factors, GRS aggregates multiple genetic variants associated with CVD to estimate a person's inherited risk [15]. PRS incorporate information from hundreds to millions of single nucleotide polymorphisms (SNPs), each contributing a small effect, to generate a cumulative risk score. These scores are stable across the lifespan and can be calculated from birth, offering opportunities for early risk assessment and intervention [16]. Studies have identified key genetic loci, such as LDLR, PCSK9, LPA, and APOE, which influence cholesterol metabolism, inflammation, and atherosclerosis progression [17]. Individuals with a high GRS have an increased lifetime risk of coronary artery disease (CAD), stroke, and other cardiovascular conditions, even in the absence of traditional risk factors like hypertension or diabetes. This suggests that PRS may offer complementary information to traditional models and improve risk stratification, especially for individuals with borderline clinical profiles or a strong family history of CVD.

Integrating genetic risk scores with conventional models enhances risk stratification, early detection, and personalized preventive strategies. For instance, research suggests that individuals with a high polygenic risk score for CAD benefit significantly from early lifestyle modifications and lipid-lowering therapies, reducing their long-term risk [18]. Furthermore, recent evidence indicates that those with a high genetic risk may achieve similar relative risk reductions from statin therapy as those with a high clinical risk, reinforcing the clinical utility of PRS. As genomic research advances, genetic risk prediction models may play a crucial role in precision medicine, helping clinicians identify high-risk individuals at an early age and implement targeted interventions for better cardiovascular health outcomes [19]. Despite this promise, barriers such as limited integration into clinical workflows, concerns about population applicability, and lack of standardization in model development and reporting remain challenges to widespread adoption. To date, there has yet to be a clear consensus on the model that should be used, and no systematic attempt to compare the key characteristics, usefulness, and limitations, as well as finding the existing gaps in the knowledge of the existing models, has been made. A comprehensive synthesis of available genetic risk models is needed to clarify their methodological differences, evaluate their comparative performance, and guide future research and clinical application.

In this systematic review, a comprehensive and up-to-date overview of the existing published genetic risk prediction models for CVD was provided. The aim was also to summarize the key features of each model such as variables used, risk stratification, and the comparative performance in terms of sensitivity, specificity, balanced accuracy, and area under the curve (AUC) values so that practitioners can make an informed choice.

2. Materials and methods

A systematic screening of articles was conducted using the Preferred Reporting Items for Systematic Reviews and Meta-Analysis (PRISMA) flow chart, and the required information was extracted using the critical appraisal and data extraction for systematic reviews of prediction modeling studies CHARMS checklist.

2.1 Search strategy

A comprehensive search strategy was developed using Boolean operators and Medical Subject Headings (MeSH terms) to ensure the retrieval of relevant studies on cardiovascular genetic risk prediction models. The search was conducted across three major databases: PubMed, Scopus, and Web of Science, with additional searches in Google Scholar for gray literature.

To maximize the sensitivity and specificity of the search, a combination of Boolean operators (AND, OR, NOT) was used. The primary search terms included "Cardiovascular Diseases" OR "Heart Disease" OR "Coronary Artery Disease" OR "CVD," combined with "Genetic Risk" OR "Polygenic Risk Score" OR "Genetic Predisposition" OR "Hereditary Factors" to identify studies focusing on genetic risk factors. Additionally, terms related to predictive modeling were included, such as "Prediction Model" OR "Risk Assessment" OR "Machine Learning" OR "Statistical Model" OR "Prognostic Model." To ensure the inclusion of relevant study designs, terms such as "Cohort Study" OR "Case-Control Study" OR "Longitudinal Study" OR "Meta-Analysis" were also applied. The Boolean operator "AND" was used to combine these distinct concepts, while "OR" was used to include synonyms and variations of terms. The "NOT" operator was applied to exclude irrelevant studies, such as those involving animal models or unrelated diseases. In PubMed, MeSH terms were integrated to refine the search further, ensuring studies were indexed according to their medical relevance. The final search query included terms such as "Cardiovascular

Diseases" [MeSH] OR "Coronary Disease" [MeSH], combined with "Genetic Predisposition to Disease" [MeSH] OR "Polygenic Risk Score" [MeSH], along with terms related to risk assessment and prognostic models. Similar structured searches were performed in Scopus and Web of Science, adjusting syntax based on the database's indexing system.

Several filters and limits were applied to enhance the quality of retrieved studies. The search was restricted to human studies, excluding animal research. Additionally, studies published from 2000 to 2024 were included to ensure the analysis was based on contemporary research. Furthermore, studies were filtered based on their design, with a focus on cohort studies, case-control studies, systematic reviews, and meta-analyses, ensuring the inclusion of high-quality evidence relevant to cardiovascular genetic risk prediction. The search process was carefully documented, and the final number of retrieved records was recorded before the screening and eligibility assessment stages. A structured search approach was used to identify studies relevant to the research question.

2.2 Eligibility criteria

All the retrieved articles were screened independently for inclusion by two researchers (S.K. and X.Z.) from this study. The titles and abstracts of retrieved papers were used as the primary review content for inclusion verification. However, if questioned or unclear, the full article was reviewed for the inclusion decision. The study's inclusion criteria include: 1. Reporting a risk assessment tool, for example, an equation or a risk score system; 2. Predicting the genetic risk of CVD; 3. Published in English-language; 4. Full text only; 5. Reporting quantitative measures of model performance (preferred but not necessarily required). Exclusion criteria include: 1. Studies that only describe an association between a single SNP and CVD; 2. Simulation studies; 3. Studies predicting (cost-) effectiveness, side effects, or developing a questionnaire; 4. Unpublished research data; 5. Studies that include risk models developed in specific patient populations such as renal disease, diabetes, atrial fibrillation, and surgery.

2.3 Selection procedure

All the retrieved articles were screened based on their titles and abstracts by a single reviewer (S.K.). Two reviewers (S.K. and X.Z.) subsequently evaluated the full text of each article independently to determine eligibility for inclusion using pre-defined criteria. In the case of disagreement between the two reviewers, a deep discussion between the two reviewers (S.K. and X.Z.) would be conducted before making a final decision about the inclusion/exclusion of the study into the review.

2.4 Data abstraction/data analysis and quality assessment

The data extraction process included two key domains: (i) Baseline characteristics of the included studies and (ii) genetic information, outcome measures, and performance evaluation. The extracted baseline characteristics comprised study authors and publication year, study design (prospective cohort, retrospective cohort, GWAS, or longitudinal study), study duration and location, population characteristics (participants' age, gender, and inclusion/exclusion criteria regarding CVD history), sample size, and source of data (e.g., UK Biobank, INTERHEART Study, EPIC-Norfolk Study, etc.).

The second domain focused on genetic data and model performance metrics. Extracted parameters included the genetic risk model used (e.g., PRS, ancestry-adjusted PRS, or multiple PRS models), method of genetic risk scoring (weighted or unweighted), number of SNPs considered, and genetic variants included (e.g., APOE, LPA, PCSK9, etc.). In addition, the confounders adjusted in the model (e.g., age, sex, LDL, BMI, smoking, diabetes) and the outcome measured (e.g., coronary heart disease, myocardial infarction, heart failure, cardiovascular disease risk) were documented.

For model evaluation, the data extracted covered follow-up duration, model validation approaches (cross-validation, split-sample, or external validation methods), and performance metrics, including indicators of overall performance, discrimination (AUC/C-statistic), and calibration (Hosmer-Lemeshow χ^2 test with p-value results). The reclassification improvement methods such as Net Reclassification Index (NRI) and Integrated Discrimination Index (IDI) were also collected where available. A formal meta-analysis of prediction models was not conducted, as this was beyond the scope of the review. For quality assessment, there is no single universally accepted framework for evaluating the quality of prediction or prognostic studies. However, in this review, the PROBAST (Prediction model Risk of Bias ASsessment Tool) was applied to assess the methodological soundness of risk prediction models [20].

3. Results

The study selection process followed the PRISMA [21] guidelines and is outlined in the PRISMA flowchart. A total of 4436 records were identified from multiple sources, including database searches from PubMed (1439), Scopus (1173), and Web of Science (990), as well as additional sources such as websites (285), organizations (150), and citation searching (399). After removing 1553 duplicate records, 2883 unique records remained for screening. During the screening phase, 2883 records were reviewed based on titles and abstracts, resulting in the exclusion of 1586 records that did not meet the inclusion criteria. The remaining 1297 records were sought for full-text retrieval, but 91 reports were not accessible, leaving 1206 full-text articles to be assessed for eligibility. Following the eligibility assessment, 728 reports were excluded for various reasons, including duplicate studies not identified earlier (n = 200), studies not meeting inclusion criteria (e.g., incorrect outcomes or predictors, n = 250), insufficient methodology details (n = 150), and lack of full-text availability (n = 128). Ultimately, nine studies were included in the final analysis. Among these, three studies focused on both development and validation, four studies were dedicated to development only, and two studies focused solely on validation. This systematic selection process ensured a rigorous and unbiased inclusion of relevant studies, allowing for a comprehensive evaluation of cardiovascular genetic risk prediction models (**Figure 1**).

3.1 Baseline characteristics of included studies

A total of nine studies were included, each varying in study design, location, participant demographics, and duration. According to the study design, seven studies followed a prospective cohort design [22–28], whereas only one study utilize the retrospective cohort approach (**Table 1**) [29]. GWAS + validation approach was

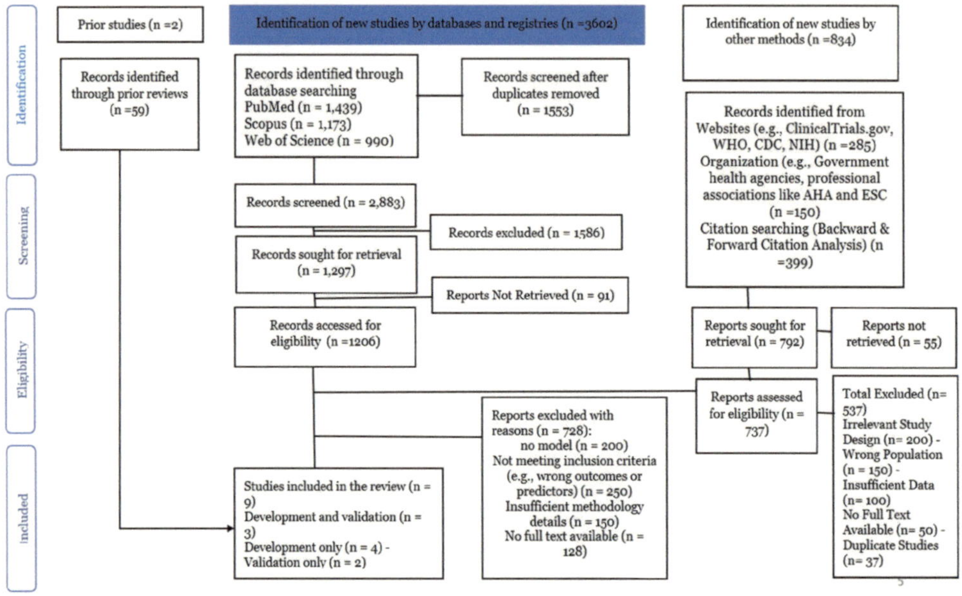

Figure 1.
PRISMA flowchart depicting the study selection process (Adopted from Ref. [21]).

reported by one study [30]. The study durations ranged from 6 to 17 years, with the longest being Park et al. [28] (2007–2024). Regarding the study location, three studies were conducted in the United Kingdom (UK Biobank: [23, 24, 26] and one study from Europe (EPIC-Norfolk Study: [22]). About the data source, two studies analyzed multi-country data (INTERHEART: [25] Global Biobank Meta-Analysis: [30]. One study targeted a multi-ethnic population (All of Us Research Program: [29], and two studies examined country-specific cohorts (Greek Population Study: [27] Korean Genome and Epidemiology Study: [28]. Most of the studies analyzed adult populations, with five studies focusing on individuals without prior CVD history [22, 23, 25, 26], and four studies included mixed populations (with and without prior CVD) [27–30]. The largest cohort was from INTERHEART (Mars et al., 2020 [25], n = 500,000), followed by UK Biobank studies (Khera et al., 2018; [23], n = 400,000; Inouye et al., 2018 [24], n = 482,500), and the smallest sample size was Patel et al. (2023) [30] (n = 29,389). Two studies ([27, 28] did not specify the participant count. The mean age of participants ranged from 49 to 53 years, with the oldest participants in Abraham et al. [22] (mean age = 53), and the youngest in Inouye et al. [24] (mean age = 49). Sex distribution was reported in seven studies, with male participants ranging from 54 to 60% ([23–26, 29, 30] and female participants ranging from 40 to 46% ([23–26, 29, 30]. Panagiotou et al. [27] and Park et al. [28] did not report sex distribution (**Table 1**).

3.2 Genetic information and outcome measured

Out of the nine studies, eight studies employed a Polygenic Risk Score (PRS)-based model ([22] [23–28, 30], and one study used an Ancestry-Adjusted PRS (**Table 2**) [30]. Seven studies used a weighted approach ([22–26, 28, 30], and two studies applied an unweighted method ([27, 29]). All studies sourced genetic markers

Study characteristics	**Source**								
	Developmental studies			Internal validation		External validation			
	Abraham et al. [22]	**Khera et al. [23]**	**Inouye et al. [24]**	**Mars et al. [25]**	**Torkamani et al. [29]**	**Gola et al. [26]**	**Patel et al. [30]**	**Panagiotou et al. [27]**	**Park et al. [28]**
Study name	EPIC-Norfolk Study	UK Biobank	UK Biobank	INTERHEART Study	All of Us Research Program	UK Biobank	Global Biobank Meta-Analysis	Greek Population Study	Korean Genome and Epidemiology Study
Study duration	2006–2016	2010–2018	2012–2018	2008–2020	2008–2018	2012–2021	2015–2024	2017–2023	2007–2024
Study design	Prospective Cohort	Prospective Cohort	Prospective Cohort	Prospective Cohort	Retrospective Cohort	Prospective Cohort	GWAS + Validation	Prospective Cohort	Longitudinal Study
Study location	Europe	United Kingdom	United Kingdom	Multi-Country	Multi-Country	United Kingdom	Multi-Country	Greece	Korea
Data/population source	Adult Population without prior CVD history	Adult Population without prior CVD history	Multi-Ethnic Adult Population with/without prior CVD history	Adult Population without prior CVD history	Adult Population with/without prior CVD history	Adult Population without prior CVD history	Multi-Country Adult Population with/without prior CVD history	Adult Population with/without prior CVD history	Adult Population with/without prior CVD history
Participant numbers	120,000	400,000	482,500	500,000	80,000	300,000	29,389	Not specified	Not specified
Mean age	53	52	49	51	50	50	50	Not specified	Not specified
Sex percentage									
Male	58	54%	59%	55%	60%	54%	54%	Not specified	Not specified
Female	42	46%	41%	45%	40%	46%	46%	Not specified	Not specified
Reference	[22]	[23]	[24]	[25]	[29]	[26]	[30]	[27]	[28]

CVD, cardiovascular disease.

Table 1.
Baseline characteristics of the included studies.

Study characteristics	Source								
	Developmental studies			Internal validation			External validation		
	Abraham et al. [22]	**Khera et al. [23]**	**Inouye et al. [24]**	**Mars et al. [25]**	**Torkamani et al. [29]**	**Gola et al. [26]**	**Patel et al. [30]**	**Panagiotou et al. [27]**	**Park et al. [28]**
Model used	PRS	PRS	PRS	PRS	PRS	PRS	Ancestry-Adjusted PRS	PRS	Multiple PRS Models
Method of GRS	Weighted	Weighted	Weighted	Weighted	Unweighted	Weighted	Weighted	Unweighted	Weighted
Markers used	Multiple	Multiple	Multiple	Multiple	Multiple	Multiple	Multiple	Multiple	Multiple
No of SNPs	30	50	1,700,000	100	200	150	250	75	300
Source of genetic markers	GWAS	GWAS	GWAS	GWAS	GWAS	GWAS	GWAS	GWAS	GWAS
Genetic variant	APOE, LPA, SCARB1	LDLR, PCSK9, LPA, APOE	LPA, CETP, FTO, TCF7L2	PCSK9, APOE, FTO	APOE, MYH7, NOS3	PCSK9, LPA, MYH7, NOS3	LPA, NOS3, PCSK9	Not specified	Not specified
Confounders Adjusted in the Model	Age, Sex, LDL, Smoking	Age, Sex, Smoking, Diabetes	Age, LDL, BMI, Physical Activity	Age, Smoking, Hypertension	Age, BMI, Diabetes, HDL	Age, LDL, BMI, Smoking	Age, Hypertension, Diabetes	Age, Smoking, HDL, LDL	Age, Physical Activity, Cholesterol
Outcome measured	Coronary heart disease	Coronary heart disease	Coronary heart disease	Coronary heart disease	CVD Risk Stratification	CAD & Myocardial Infarction	Heart Failure & Stroke	Coronary Artery Disease	Cardiovascular Disease Risk
Follow up (years)	8	10	15	12	NA	NA	NA	6	17
Model performance measures									
Calibration									
Method	Hosmer-Lemeshow	Chi-square	Hosmer-Lemeshow	Hosmer-Lemeshow	NA	Chi-square	Hosmer-Lemeshow	NA	NA

Study characteristics	Source								
	Developmental studies			Internal validation		External validation			
	Abraham et al. [22]	**Khera et al. [23]**	**Inouye et al. [24]**	**Mars et al. [25]**	**Torkamani et al. [29]**	**Gola et al. [26]**	**Patel et al. [30]**	**Panagiotou et al. [27]**	**Park et al. [28]**
Result (p)	0.02	0.01	0.03	0.04	NA	0.02	0.01	NA	NA
Discrimination									
Method	AUROC	AUROC	AUROC	AUROC	AUROC	AUROC	AUROC	AUROC	AUROC
Result (p)	0.76	0.81	0.83	0.83	0.75	0.80	0.78	0.72	0.74
Validation mode	Cross-validation	Cross-validation	Split-sample	Cross-validation	Cross-validation	Split-sample	Cross-validation	Cross-validation	Split-sample
Reclassification (NRI/IDI)	NRI = 0.10	IDI = 0.05	NRI = 0.12	NRI = 0.08	NA	IDI = 0.04	NRI = 0.09	NA	NA
Evidence of clinical utility	Minimal	Yes	Yes	Yes	Minimal	Yes	Minimal	No	Minimal
Comments	Needs further validation in diverse populations	Shows improvement over conventional risk models	Supports integration into clinical practice	Potential for risk stratification in primary care	Requires validation in larger cohorts	Demonstrates potential for clinical implementation	Requires further studies on generalizability	No clinical implementation evidence provided	Clinical relevance uncertain, requires future trials

APOE, Apolipoprotein E; AUROC, Area Under the Receiver Operating Characteristic Curve; BMI, Body Mass Index; DBP, Diastolic Blood Pressure; FTO, Fat Mass and Obesity-Associated Gene; GWAS, Genome-Wide Association Study; HDL, High-Density Lipoprotein; IDI, Integrated Discrimination Index; LDL, Low-Density Lipoprotein; LDLR, Low-Density Lipoprotein Receptor; LPA, Lipoprotein (a); MYH7, Myosin Heavy Chain 7; NA, Not Available/Not Applicable; NRI, Net Reclassification Index; NOS3, Nitric Oxide Synthase 3; PCSK9, Proprotein Convertase Subtilisin/Kexin Type 9; PRS, Polygenic Risk Score; SBP, Systolic Blood Pressure; TCF7L2, Transcription Factor 7 Like 2; TG, Triglycerides.

Table 2.
Genetic information, outcome measured, and performance evaluation of the included studies.

from GWAS. The number of SNPs (Single Nucleotide Polymorphisms) varied, with the highest being 1,700,000 [24] and the lowest being 30 [22].

The most frequently included genetic variants across the studies were APOE [22, 23, 25, 29], LPA [22–24, 26, 30], PCSK9 ([23, 25, 26, 30], FTO [24, 25]), MYH7 [26, 29], NOS3 [26, 29, 30], and the least ones were SCARB1 [22], LDLR (Khera et al., 2018), CETP [24], and TCF7L2 [24]. The genetic markers were not specified for Panagiotou et al. [27] and Park et al. [28] in the extracted data.

The most common adjusted confounder in the models was age, which was included in all studies (**Table 2**) (n = 9; 100%). Sex was adjusted for in five studies (, LDL cholesterol was considered in five studies (, and BMI was included as a confounder in four studies . Smoking was considered in four studies (, hypertension was included in three studies , and diabetes was included in four studies .

Coronary heart disease (CHD) was the most common outcome, assessed in five studies [22–25, 28], whereas two studies evaluated CAD and myocardial infarction (**Table 2**) [26, 30]. On the other hand, only one study focused on heart failure and stroke ([27], and one study measured general cardiovascular disease risk [28]. The follow-up duration ranged from 6 years [27] to 17 years [28].

3.3 Model performance evaluation

For performance evaluation, calibration was assessed using the Hosmer–Lemeshow test in five studies [22, 24, 25, 30], while two studies [23, 26] applied the Chi-square test. However, Park et al. [27, 28] did not report any calibration assessment (**Table 2**), while discrimination was assessed using AUROC scores, which ranged from 0.72 to 0.83 across the studies. The highest AUROC was observed in Inouye et al. (0.83) and Mars et al. [25] (0.83), while the lowest AUROC was reported in Panagiotou et al. [27] (0.72).

Additionally, among the validation approaches, cross-validation was the most frequently used method, applied in five studies [23–25, 28, 30], while split-sample validation was employed in three studies (**Table 2**) [25, 27, 28]. Additionally, external validation was performed in two studies [22, 26]. Regarding reclassification metrics, the Net Reclassification Index (NRI) was reported in four studies [22, 24, 26, 28], whereas the Integrated Discrimination Index (IDI) was documented in two studies [23, 27].

Lastly, the evidence of clinical utility suggests that four studies demonstrated potential for clinical implementation (**Table 2**) [23–25, 30]. Three studies required further validation ([26–28], and two studies had minimal clinical evidence ([22, 29]).

3.4 Quality assessment of the included studies

The risk of bias for the nine included studies was assessed using PROBAST (Prediction Model Risk of Bias Assessment Tool) [20], which evaluates four key domains: participants, predictors, outcome, and analysis. The overall risk of bias was categorized as low, moderate, or high based on methodological rigor and potential biases in study design. Among the included studies, three studies [24, 26, 28] had a low risk of bias across all domains (**Table 3, Figure 2**). These studies used well-defined participant selection criteria, robust predictor measurements, clearly defined outcomes, and appropriate statistical methods, ensuring high methodological reliability. Four studies [22, 23, 25, 30] were assessed as having a moderate risk of bias. The primary concerns in these studies were related to outcome assessment and statistical

Study type	Study name/study source	Participants	Predictors	Outcome	Analysis	Overall risk of bias
Developmental	Abraham et al. [22]	🟢	🟢	🟡	🟡	🟡
	Khera et al. [23]	🟢	🟢	🟢	🔴	🟡
	Inouye et al. [24]	🟢	🟢	🟢	🟢	🟢
Internal validation	Mars et al. [25]	🟡	🟢	🔴	🟡	🟡
	Torkamani et al. [29]	🔴	🟡	🔴	🔴	🔴
External validation	Gola et al. [26]	🟢	🟢	🟢	🟢	🟢
	Patel et al. [30]	🟢	🟡	🟢	🟡	🟡
	Panagiotou et al. [27]	🔴	🟡	🔴	🔴	🔴
	Park et al. [28]	🟢	🟢	🟢	🟢	🟢

Note: 🟢 = *Low risk,* 🟡 = *Moderate risk,* 🔴 = *High risk.*

Table 3.
Quality assessment of the included studies using PROBLAST [20].

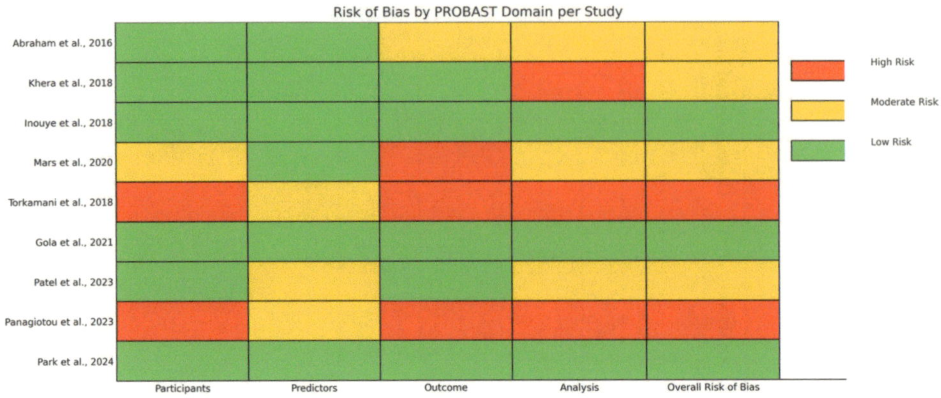

Figure 2.
Quality assessment of the included studies using PROBLAST [20].

analysis. For example, Abraham et al. [22] and Mars et al. [25] had moderate risk in outcome and analysis domains, indicating potential measurement inconsistencies or methodological limitations. Khera et al. [23] demonstrated high risk in analysis, suggesting possible overfitting or lack of external validation (**Table 3**, **Figure 2**).

4. Discussion

This review assessed the methodological quality and performance of various cardiovascular disease (CVD) prediction models, focusing on their study characteristics, genetic risk factors, performance metrics, and validation approaches. The findings highlight variations in study design, population selection, and statistical methodologies, influencing the overall reliability and usability of these models.

The included studies employed prospective, retrospective, and genetic association-based designs, reflecting differences in data collection and follow-up methods. While prospective cohort studies [23, 24, 26, 28] provided more robust long-term insights, retrospective analyses [29] relied on previously collected data, introducing potential selection bias. Additionally, multi-country studies [25, 30] increased the generalizability of findings, whereas single-center studies [27, 28] may be less applicable across diverse populations. The variability in sample size and participant demographics also affects model generalizability. Larger datasets, such as INTERHEART (Mars et al., 2020, n = 500,000) and UK Biobank studies (Khera et al., 2018, n = 400,000; Inouye et al., 2018, n = 482,500), provide statistically powerful insights, whereas smaller cohorts (Patel et al. [30], n = 29,389) may lack sufficient representation for risk prediction in broader populations. Furthermore, some studies did not specify sex distribution [28], limiting the ability to analyze sex-based variations in CVD risk.

A key aspect of this review was the integration of genetic predictors into risk assessment models. The majority of studies incorporated Polygenic Risk Scores (PRS), with variations in scoring methodologies (weighted vs. unweighted) and the number of single nucleotide polymorphisms (SNPs) considered. The most frequently included genetic variants were APOE, LPA, and PCSK9, which have well-established links to lipid metabolism and cardiovascular risk ([22, 23, 25, 26, 30]. However, not all studies utilized the same set of genetic markers, potentially affecting predictive consistency across populations. Additionally, clinical and lifestyle factors such as LDL cholesterol, BMI, diabetes, and smoking were included as confounders in different models. However, the lack of standardized predictor inclusion across studies raises concerns about model comparability. For example, while LDL cholesterol was adjusted for in five studies [22, 24–27], other relevant biomarkers, such as HDL cholesterol and triglycerides, were not consistently reported [26]. This inconsistency may limit the reproducibility and clinical applicability of certain models.

Regarding performance evaluation, discrimination ability of the models, assessed through AUROC (C-statistic), ranged from 0.72 to 0.83. Notably, Inouye et al. and Mars et al. [25] achieved the highest AUROC values (0.83), indicating strong predictive performance. In contrast, Panagiotou et al. [27] reported the lowest AUROC (0.72), suggesting room for improvement in model accuracy. Calibration was assessed through Hosmer–Lemeshow tests in five studies ([22, 24, 25, 30] and Chi-square tests in two studies [23, 26]. However, Park et al. [28] and Panagiotou et al. [27] did not report calibration results, making it difficult to assess model reliability in those cases. For validation approaches, cross-validation was the most common method (used in five studies), whereas split-sample and external validation were less frequently applied. The lack of extensive external validation in most models suggests potential limitations in their applicability across diverse populations. Furthermore, reclassification metrics such as Net Reclassification Index (NRI) and Integrated Discrimination Index (IDI) were only reported in a few studies, highlighting the need for better performance evaluation measures.

Quality assessment using PROBAST [20] revealed that most studies had a low or unclear risk of bias, particularly in predictor and outcome assessments. However, Torkamani et al. [29] and Gola et al. [26] were identified as high risk, primarily due to concerns regarding participant selection and statistical analysis methods. Additionally, some studies lacked key methodological details, such as missing sex distribution data [27, 28] or not reporting calibration results, which limits the transparency and reproducibility of findings. Moreover, while some models demonstrated clinical applicability ([23–25, 30], others lacked sufficient validation to be integrated into clinical practice.

The findings suggest that PRS-based models show promise in enhancing cardiovascular risk prediction, particularly when combined with traditional clinical factors. However, standardization in predictor selection, validation techniques, and performance evaluation methods is needed to improve model comparability. Future studies should focus on (1) external validation in diverse populations to enhance generalizability, (2) incorporating additional clinical markers to improve model performance, and (3) reporting detailed calibration and reclassification metrics to better assess clinical utility.

5. Strengths and limitations

This systematic review offers several key strengths. To our knowledge, it is the first review to focus exclusively on genetic risk prediction models for cardiovascular disease (CVD), providing a comprehensive and up-to-date synthesis of published models. The study employed a rigorous and transparent methodology, adhering to CHARMS checklist and utilizing the PROBAST tool to assess the risk of bias across the included studies. By systematically evaluating the core features of each model—including study design, population, predictors, genetic variants, statistical approaches, and performance metrics—this review offers practical insights for clinicians, researchers, and policymakers aiming to integrate genetic risk into cardiovascular prevention strategies. The findings also highlight current gaps in model validation, standardization, and applicability, especially in diverse or underrepresented populations, thereby offering a roadmap for future research in the field.

However, the review also has several limitations. Despite employing a comprehensive search strategy, it is possible that some relevant studies were missed, particularly unpublished models or studies not indexed in major English-language databases. The substantial heterogeneity among the included studies in terms of design, outcome definitions, and predictor sets precluded a formal meta-analysis or pooled quantitative synthesis. Moreover, most models were developed and validated in European or North American populations, which may limit their generalizability to other regions, particularly low- and middle-income countries. Additionally, reliance on published data meant that some studies lacked important methodological details, such as calibration metrics or external validation outcomes, which may affect the accuracy of our comparative assessments. Finally, given the rapid pace of genomic research, newer models may have been published since the review's search cut-off, underscoring the need for periodic updates to maintain relevance.

6. Conclusion

In conclusion, this systematic review highlights the increasing importance of genetic risk prediction models in improving cardiovascular disease (CVD) risk assessment. Polygenic risk scores (PRS) offer a promising complement to traditional clinical models, enabling the earlier and more individualized identification of high-risk individuals. However, despite these advancements, significant challenges persist—most notably, inconsistencies in study design, heterogeneity in predictor selection, and limited external validation across diverse populations. These limitations hinder the clinical translation and generalizability of existing models. Addressing these gaps through standardized methodologies, diverse population-based validations, and integration with clinical data will be essential for developing robust, reliable, and equitable risk prediction tools. Ultimately, refining these models will support more personalized preventive strategies and make a meaningful contribution to the goals of precision cardiovascular medicine.

Funding sources

Nil.

Disclosures

None declared.

Author details

Sumera Khan[1*] and Xianquan Zhan[2]

1 School of Pharmaceutical Sciences, Comsats University Islamabad, Pakistan

2 Shandong First Medical University, China

*Address all correspondence to: khansumeera@gmail.com

References

[1] Luo Y et al. Global burden of cardiovascular diseases attributed to low physical activity: An analysis of 204 countries and territories between 1990 and 2019. American Journal of Preventive Cardiology. 2024;**17**:100633

[2] Wang Y et al. Global trends in the burden of ischemic heart disease based on the global burden of disease study 2021: The role of metabolic risk factors. BMC Public Health. 2025;**25**(1):310

[3] Anand S, Bradshaw C, Prabhakaran D. Prevention and management of CVD in LMICs: Why do ethnicity, culture, and context matter? BMC Medicine. 2020;**18**(1):7

[4] D'Agostino RB et al. General cardiovascular risk profile for use in primary care. Circulation. 2008;**117**(6):743-753

[5] Ridker PM et al. Development and validation of improved algorithms for the assessment of global cardiovascular risk in women the Reynolds risk score. JAMA. 2007;**297**(6):611-619

[6] Ridker PM et al. C-reactive protein and parental history improve global cardiovascular risk prediction: The Reynolds risk score for men. Circulation. 2008;**118**(22):2243-2251

[7] Conroy RM et al. Estimation of ten-year risk of fatal cardiovascular disease in Europe: The SCORE project. European Heart Journal. 2003;**24**(11):987-1003

[8] Assmann G, Cullen P, Schulte H. Simple scoring scheme for calculating the risk of acute coronary events based on the 10-year follow-up of the prospective cardiovascular Münster (PROCAM) study. Circulation. 2002;**105**(3):310-315

[9] Goff DC et al. 2013 ACC/AHA guideline on the assessment of cardiovascular risk. Circulation. 2014;**129**(25_suppl. 2):S49-S73

[10] Collins GS, Altman DG. An independent and external validation of QRISK2 cardiovascular disease risk score: A prospective open cohort study. BMJ. 2010;**340**:c2442

[11] JBS3 Board. Joint British societies' consensus recommendations for the prevention of cardiovascular disease (JBS3). Heart. 2014;**100 Suppl 2**:ii1-ii67

[12] Xu Z et al. Prediction of cardiovascular disease risk accounting for future initiation of statin treatment. American Journal of Epidemiology. 2021;**190**(10):2000-2014

[13] Upadhyay RK. Emerging risk biomarkers in cardiovascular diseases and disorders. Journal of Lipids. 2015;**2015**:971453

[14] Knowles JW, Ashley EA. Cardiovascular disease: The rise of the genetic risk score. PLoS Medicine. 2018;**15**(3):e1002546

[15] Xiang R et al. Recent advances in polygenic scores: Translation, equitability, methods and FAIR tools. Genome Medicine. 2024;**16**(1):33

[16] Abraham G, Rutten-Jacobs L, Inouye M. Risk prediction using polygenic risk scores for prevention of stroke and other cardiovascular diseases. Stroke. 2021;**52**(9):2983-2991

[17] Paththinige CS, Sirisena ND, Dissanayake VHW. Genetic determinants of inherited susceptibility to hypercholesterolemia – A comprehensive

literature review. Lipids in Health and Disease. 2017;**16**(1):103

[18] Klarin D, Natarajan P. Clinical utility of polygenic risk scores for coronary artery disease. Nature Reviews. Cardiology. 2022;**19**(5):291-301

[19] O'Sullivan JW et al. Polygenic risk scores for cardiovascular disease: A scientific statement from the American Heart Association. Circulation. 2022;**146**(8):e93-e118

[20] Wolff RF et al. PROBAST: A tool to assess the risk of bias and applicability of prediction model studies. Annals of Internal Medicine. 2019;**170**(1):51-58

[21] Page MJ et al. PRISMA 2020 explanation and elaboration: Updated guidance and exemplars for reporting systematic reviews. BMJ. 2021;**372**:n160

[22] Abraham G et al. Genomic prediction of coronary heart disease. European Heart Journal. 2016;**37**(43):3267-3278

[23] Khera AV et al. Genome-wide polygenic scores for common diseases identify individuals with risk equivalent to monogenic mutations. Nature Genetics. 2018;**50**(9):1219-1224

[24] Inouye M et al. Genomic risk prediction of coronary artery disease in 480,000 adults: Implications for primary prevention. Journal of the American College of Cardiology. 2018;**72**(16):1883-1893

[25] Mars N et al. Polygenic and clinical risk scores and their impact on age at onset and prediction of cardiometabolic diseases and common cancers. Nature Medicine. 2020;**26**(4):549-557

[26] Gola D et al. Population bias in polygenic risk prediction models for coronary artery disease. Circulation: Genomic and Precision Medicine. 2020;**13**(6):e002932

[27] Panagiotou N, Chatziandreou E. Adjusted polygenic risk score enables personalized cardiovascular disease prevention and clinical management. Value in Health. 2022;**25**(12):S378-S379

[28] Park YS et al. Evaluating cardiovascular disease risk stratification using multiple-polygenic risk scores and pooled cohort equations: Insights from a 17-year longitudinal Korean cohort study. Frontiers in Genetics. 2024;**15**:1364993

[29] Torkamani A, Wineinger NE, Topol EJ. The personal and clinical utility of polygenic risk scores. Nature Reviews. Genetics. 2018;**19**(9):581-590

[30] Patel AP et al. A multi-ancestry polygenic risk score improves risk prediction for coronary artery disease. Nature Medicine. 2023;**29**(7):1793-1803

Chapter 4

Telomere Dynamics in Cancer: Novel Insights and Therapeutic Implications

Nosipho Treasure Mdluli and Yun-Chen Chang

Abstract

Telomere dynamics are pivotal in cancer biology, representing a balance between genomic stability and tumor progression. This chapter explores the intricate mechanisms governing telomere maintenance, including telomerase activation and the alternative lengthening telomeres (ALT) pathway, enabling cancer cells to bypass replicative senescence. It highlights recent advances in understanding the molecular underpinnings of telomere regulation and their implications for diagnostics and therapeutics. The chapter also examines telomere-targeting strategies, such as telomerase inhibitors and ALT-specific interventions, and their potential for clinical translation. By addressing emerging research and therapeutic innovations, this chapter underscores the critical role of telomeres in precision oncology and personalized medicine.

Keywords: telomere maintenance, cancer progression, telomerase activation, alternative lengthening of telomeres (ALT), therapeutic targets

1. Introduction

Telomeres, the protective caps of chromosomes composed of repetitive TTAGGG sequences, play a crucial role in safeguarding genomic integrity by preventing chromosome ends from being misrecognized as double-strand breaks. These specialized structures and the associated shelterin protein complex maintain chromosomal stability and regulate cellular lifespan. However, due to the "end replication problem," telomeres naturally shorten with each cell division. This gradual attrition serves as a molecular clock that limits the replicative potential of somatic cells, ultimately inducing replicative senescence or apoptosis when telomeres become critically short [1, 2].

While this process acts as a tumor-suppressive mechanism, cancer cells circumvent this limitation to achieve limitless replicative capacity—a hallmark of cancer. To overcome telomere shortening, most tumors reactivate telomerase, a reverse transcriptase that elongates telomeres and is normally inactive in somatic cells. Approximately 85–90% of human cancers exploit telomerase reactivation to support continuous cell division [3, 4]. The remaining cancers rely on the alternative lengthening of telomeres (ALT), a recombination-based mechanism that maintains telomere length through homologous recombination [5]. Both telomerase and

IntechOpen

ALT mechanisms allow cancer cells to bypass senescence and apoptosis, enabling unchecked growth and metastasis.

Recent research has revealed that telomeres not merely are passive chromosomal end protectors but actively participate in cellular processes that influence cancer progression. Telomere dysfunction caused by excessive shortening or deproteination of shelterin components can trigger DNA damage responses, genomic instability, and chromosomal rearrangements [6, 7]. Additionally, the noncanonical roles of telomerase in promoting oncogenic signaling, inflammation, and stemness are emerging as critical areas of investigation [8].

Beyond their biological significance, telomeres and telomerase are promising clinical tools. Short telomeres in peripheral blood leukocytes have been associated with increased cancer risk and aggressive disease phenotypes, suggesting their potential as biomarkers for early detection and prognosis [9, 10]. Telomerase activity and ALT signatures also provide opportunities for personalized therapy, as they represent cancer-specific vulnerabilities that can be exploited with targeted interventions.

Therapeutic strategies targeting telomere maintenance mechanisms are undergoing rapid development. Telomerase inhibitors, such as imetelstat, have shown promise in preclinical and clinical studies by inducing telomere shortening and senescence in cancer cells [11]. Meanwhile, ALT-specific therapies targeting DNA repair enzymes, such as ATR inhibitors, are being investigated for their efficacy against ALT-positive tumors [12]. These approaches highlight the translational potential of telomere biology in oncology and underscore the importance of continued research to overcome challenges such as resistance and off-target effects.

This chapter provides a comprehensive overview of telomere dynamics in cancer, focusing on their mechanistic roles, clinical implications, and therapeutic potential. By integrating recent advances and identifying gaps in current knowledge, it aims to foster a deeper understanding of telomere biology and its relevance in precision medicine.

2. The biology of telomeres and telomerase

2.1 Telomere structure and function

Telomeres are specialized structures located at the ends of eukaryotic chromosomes, composed of repetitive nucleotide sequences (TTAGGG repeats in vertebrates) and a specialized protein complex known as shelterin. This unique composition ensures the protection and stability of the chromosome ends. The repetitive DNA sequences serve as a buffer zone that absorbs the inevitable loss of nucleotides during DNA replication, while the shelterin complex protects telomeres from being recognized as DNA double-strand breaks by the cellular repair machinery (**Figure 1**) [1, 13].

The shelterin complex comprises six proteins: TRF1, TRF2, POT1, TIN2, RAP1, and TPP1. TRF1 and TRF2 bind directly to the double-stranded telomeric DNA, while POT1 interacts with the single-stranded overhang. Together, these proteins prevent the activation of DNA damage response pathways that could otherwise lead to deleterious outcomes such as end-to-end chromosome fusions and genomic instability [14]. Shelterin also regulates the formation of T-loops, a structure in which the single-stranded overhang loops back to invade the double-stranded telomeric DNA. This configuration physically conceals the chromosome ends, further preventing inappropriate recognition and degradation [5, 15].

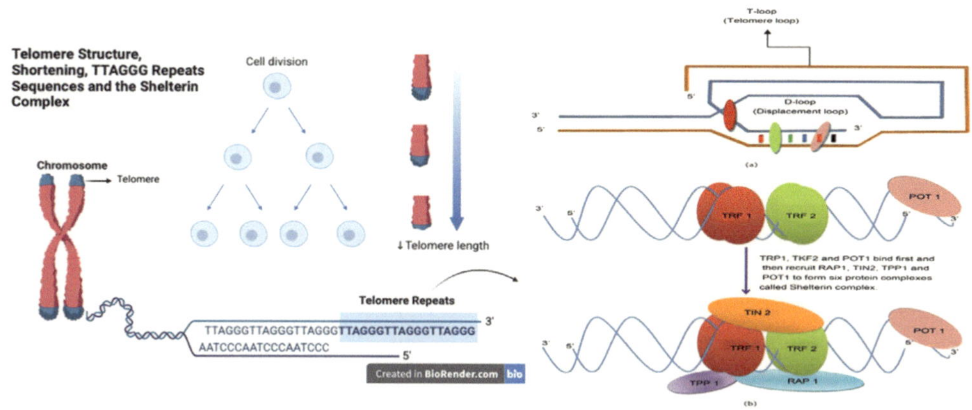

Figure 1.
Telomere structure, shortening, TTAGGG repeats sequences, and the shelterin complex.

Telomeres play a vital role in maintaining genomic stability. Dysfunctional telomeres, caused by loss of shelterin or excessive shortening, can lead to chromosomal abnormalities such as translocations, aneuploidy, and breakage-fusion-bridge cycles. These events drive genetic instability, a hallmark of cancer progression [16].

2.2 Telomere shortening and cellular senescence

During DNA replication, the inability of DNA polymerase to fully replicate the 3′ ends of linear chromosomes results in gradual telomere shortening, a phenomenon known as the "end replication problem" [2]. Each cell division removes 50–200 base pairs of telomeric DNA, eventually leading to critically short telomeres incapable of maintaining protective function. When telomeres reach this critical length, they activate DNA damage responses (DDR), mediated by ATM and ATR kinases, which recognize uncapped telomeres as double-strand breaks (**Figure 2**) [17].

This DDR induces a state of permanent cell cycle arrest known as replicative senescence. Senescence is a tumor-suppressive mechanism that prevents the uncontrolled proliferation of damaged or aged cells. However, it also contributes to aging

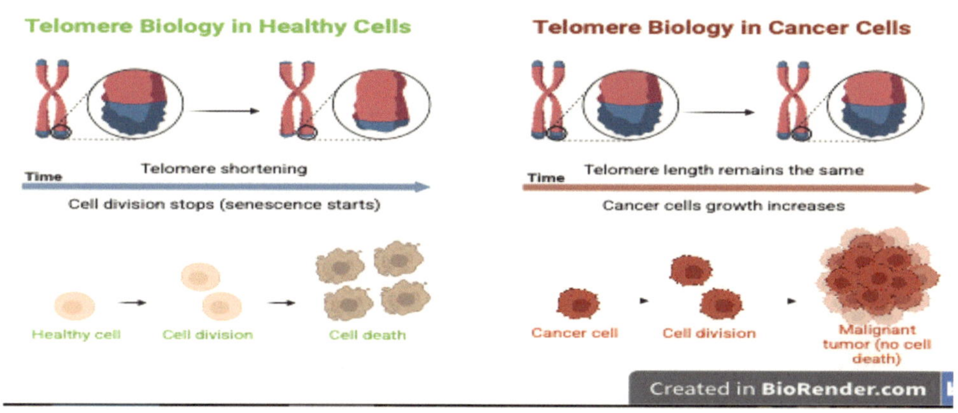

Figure 2.
Mechanism of telomere shortening.

and tissue dysfunction as senescent cells accumulate over time and secrete inflammatory cytokines and matrix-degrading enzymes, collectively called the senescence-associated secretory phenotype (SASP) [18, 19].

In certain contexts, particularly when the DDR pathways are impaired, critically short telomeres may trigger genomic instability instead of senescence, promoting cellular transformation and tumorigenesis [8].

2.3 Telomerase: The key to immortality in cancer cells

Telomerase is a ribonucleoprotein enzyme complex that counters telomere shortening by adding TTAGGG repeats to the ends of chromosomes, thereby restoring telomere length and enabling continued cellular proliferation. The core components of telomerase include:

- *TERT (Telomerase reverse transcriptase)*: The catalytic subunit responsible for synthesizing telomeric DNA from an RNA template.

- *TERC (Telomerase RNA component)*: The RNA subunit is the telomeric repeat synthesis template [4].

Telomerase is active in embryonic stem cells and certain progenitor cells but is repressed in most somatic cells during development. This repression ensures that somatic cells undergo replicative aging, a mechanism of tumor suppression. However, telomerase reactivation is a near-universal feature of cancer, observed in approximately 85–90% of tumors [3].

2.4 Mechanisms of telomerase activation in cancer telomerase activation in cancer cells occurs through several mechanisms, including

1. *TERT promoter mutations:* Mutations in the TERT promoter region are frequently observed in cancers such as glioblastomas, melanomas, and hepatocellular carcinomas. These mutations create binding sites for transcription factors that drive TERT overexpression (**Figure 3**) [20, 21].

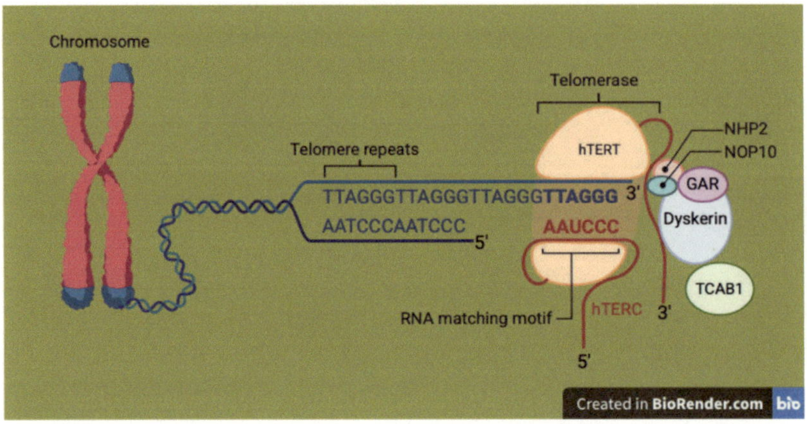

Figure 3.
Mechanism of telomerase activation.

2. *Chromosomal rearrangements:* Structural changes such as gene amplifications or rearrangements near the TERT locus can increase its transcriptional activity [20].

3. *Epigenetic modifications:* Alterations in DNA methylation and histone acetylation in the TERT promoter region can relieve transcriptional repression, promoting telomerase activity [21].

4. *Activation by oncogenic pathways*: Certain oncogenic signals, such as those mediated by c-Myc, can directly upregulate TERT expression [22].

While telomerase reactivation is a defining feature of most cancers, a subset of malignancies, particularly sarcomas and gliomas, maintain telomere length through the alternative lengthening of telomeres (ALT). ALT is a telomerase-independent mechanism that relies on homologous recombination and is associated with high levels of genomic instability [23].

The dual roles of telomerase and ALT in maintaining telomeres highlight their significance in cancer biology and their potential as therapeutic targets. By unraveling the complex regulatory mechanisms underlying telomere maintenance, researchers are uncovering new opportunities for targeted interventions in cancer therapy.

2.5 Telomere maintenance mechanisms in cancer

Cancer cells rely on specific mechanisms to maintain telomere length and ensure their ability to proliferate indefinitely. Telomere maintenance is critical to bypassing replicative senescence and crisis, enabling tumor progression and metastasis. The two primary pathways for telomere maintenance in cancer are telomerase activation and the alternative lengthening of telomeres (ALT).

2.6 Telomerase activation in tumors

Telomerase activation is the predominant telomere maintenance mechanism in approximately 85–90% of human cancers [24]. Cancer cells, unlike normal somatic cells exhibiting low or absent telomerase activity, reactivate telomerase to prevent telomere shortening and circumvent senescence or apoptosis (**Table 1**).

1. Prevalence across cancer types

   Telomerase activation is observed in a wide range of malignancies, including lung, breast, prostate, liver, and colorectal cancers. Certain tumor types, such as glioblastomas and melanomas, show particularly high frequencies of telomerase activation due to TERT promoter mutations [25, 26]. These mutations create binding sites for transcription factors like ETS/TCF, resulting in increased TERT expression. Other mechanisms include gene amplification, epigenetic modifications, and activation of oncogenic pathways like MYC, which upregulates TERT transcription [25, 26].

2. Role in bypassing senescence

   Telomerase extends the replicative lifespan of cancer cells by adding TTAGGG repeats to the ends of telomeres, effectively reversing the shortening that occurs

Cancer Type	Prevalence of TERT Promoter Mutations (%)	Other Mechanisms	Notes
Glioblastoma	~83%	ATRX/DAXX mutations, alternative lengthening of telomeres (ALT)	High prevalence of TERT promoter mutations, with some cases showing ALT pathways.
Oligodendroglioma	~95%	IDH mutations, co-deletion of 1p/19q chromosomes	TERT promoter mutations are prevalent, often with co-occurring IDH mutations.
Melanoma	~71%	BRAF mutations, NRAS mutations	Significant occurrence of TERT promoter mutations, often alongside BRAF or NRAS mutations.
Bladder Cancer	~66%	FGFR3 mutations, HRAS mutations	High frequency of TERT promoter mutations, frequently with FGFR3 or HRAS mutations.
Hepatocellular Carcinoma	~47%	HBV/HCV infections, p53 mutations	Notable presence of TERT promoter mutations, often associated with viral infections or p53 mutations.
Head and Neck Squamous Cell Carcinoma (HNSCC)	~21%	HPV infection, p16INK4a inactivation	TERT promoter mutations are mainly found in oral cavity cancers, often alongside HPV infection.
Papillary Thyroid Carcinoma	~10%	BRAF mutations, RAS mutations	TERT promoter mutations are present, with higher rates in less differentiated subtypes.
Non-Small Cell Lung Cancer (NSCLC)	~15%	KRAS mutations, EGFR mutations	TERT promoter mutations occur in a minority of cases, often with KRAS or EGFR mutations.
Breast Cancer	~5%	BRCA1/2 mutations, p53 mutations, HER2 amplification	TERT promoter mutations are rare, with other mechanisms like BRCA1/2 mutations more common.
Colorectal Cancer	~20%	APC mutations, mismatch repair deficiency (dMMR)	TERT promoter mutations are observed, often with APC mutations or dMMR.
Pancreatic Cancer	~25%	KRAS mutations, CDKN2A inactivation	TERT promoter mutations are present, frequently with KRAS mutations.

Table 1.
The prevalence of telomerase activation mechanisms, including TERT promoter mutations and other mechanisms, across different cancer types.

during cell division. By maintaining telomere length, telomerase allows cancer cells to evade replicative senescence, a natural tumor-suppressive mechanism [27].

3. Implications for tumor growth, metastasis, and therapy resistance

Telomerase activity is strongly associated with sustained tumor growth and metastatic potential. It enables cancer cells to divide indefinitely, promoting the formation of large tumor masses. Additionally, telomerase confers resistance to therapies that induce DNA damage, as maintained telomeres prevent telomere-driven genome instability.

- *Metastasis*: Studies suggest that telomerase plays a role in epithelial-to-mesenchymal transition (EMT), a key process in metastasis [28].
- *Therapy resistance*: Cancer cells with active telomerase exhibit reduced sensitivity to chemotherapeutic agents and radiation due to enhanced DNA repair capacity and telomere stabilization [29].

Given its central role in cancer biology, telomerase is an attractive therapeutic target. Telomerase inhibitors, such as imetelstat, and immunotherapies targeting TERT-derived peptides are under investigation for their potential to suppress tumor growth and overcome therapy resistance [30].

2.7 Alternative lengthening of telomeres (ALT)

A subset of cancers, particularly sarcomas and gliomas, maintains telomeres through the ALT pathway, an alternative, telomerase-independent mechanism. ALT is estimated to occur in 10–15% of cancers and is often associated with tumors that lack TERT promoter mutations (**Table 2**) [31, 32].

1. *ALT pathway mechanisms*

   The ALT pathway relies on homologous recombination-mediated elongation of telomeres. In ALT-positive cells:

   - *Telomere sister chromatid exchange (T-SCE)*: Telomeres from one chromosome act as templates for the elongation of telomeres on another, creating variability in telomere length [33].
   - *Break-induced replication (BIR)*: A DNA repair mechanism involved in replicating damaged telomeres contributes to their lengthening [34].
   - *C circle formation*: ALT cells produce extrachromosomal, circular telomeric DNA known as C circles, which are a hallmark of ALT activity and play a role in telomere maintenance [23].

2. *Characteristics of ALT-positive tumors*

   ALT-positive tumors exhibit distinct features, including:

   - *Telomere length heterogeneity*: ALT-positive cells have highly variable telomere lengths, ranging from very short to excessively long. This heterogeneity can be visualized using fluorescence in situ hybridization (FISH) [35].
   - *ALT-associated PML bodies (APBs)*: ALT cells display nuclear structures called ALT-associated promyelocytic leukemia (PML) bodies. These contain telomeric DNA, shelterin components, and DNA repair proteins, serving as sites for telomere elongation [36].

Cancer Type	ALT Positivity (%)	Genomic Instability	Variability in Telomere Length	Presence of ALT-Associated PML Bodies (APBs)	Notes
Glioblastoma	~10–15%	High	High	Present	ALT positivity is often associated with ATRX mutations and the presence of APBs.
Osteosarcoma	~50%	High	High	Present	High prevalence of ALT, frequently showing APBs.
Leiomyosarcoma	~30–40%	Moderate to High	High	Present	ALT positivity is common, with APBs detected in most cases.
Neuroblastoma	~20–30%	High	High	Present	ALT is observed in a subset of high-risk neuroblastomas, often with APBs.
Pancreatic Neuroendocrine Tumors	~10–25%	Moderate	High	Present	ALT positivity linked with ATRX/DAXX mutations and APBs presence.
Soft Tissue Sarcomas	~25–60%	High	High	Present	High prevalence of ALT, particularly in sarcomas with ATRX/DAXX mutations.
Lipoma	~5–10%	Low	Low	Present	ALT positivity is rare but can occur, with APBs detected in some cases

Table 2.
Key characteristics of ALT-positive cancers.

- *Increased genome instability*: While ALT extends the replicative potential of cancer cells, it also introduces chromosomal aberrations, making ALT-positive tumors genetically unstable [31].

3. *Clinical significance and detection methods*

The presence of ALT has significant clinical implications. ALT-positive tumors often exhibit resistance to telomerase inhibitors, necessitating the development of alternative therapeutic strategies. However, ALT-associated genome instability may make these tumors more vulnerable to DNA damage-inducing therapies [31, 37].

- Detection methods: ALT activity can be identified using various techniques:

- C circle assay: Measures the presence of C circles in cell extracts.

- FISH: Detects heterogeneous telomere lengths and APBs.

- Immunostaining: Identifies PML bodies associated with telomeric DNA [38–41].

Emerging therapies targeting ALT-positive tumors include inhibitors of DNA repair pathways essential for ALT function, such as the ATR and PARP pathways. These approaches aim to exploit the inherent vulnerabilities of ALT-positive cells [42–44].

3. Telomeres as biomarkers in cancer

The dynamic nature of telomeres and their critical role in genomic stability and cellular aging make them promising biomarkers in cancer. Telomere length and associated dynamics offer potential utility in diagnosis, prognosis, and therapeutic monitoring. However, translating this promise into clinical practice requires overcoming several technical and methodological challenges (**Figure 4**).

4. Telomere length as a diagnostic tool

Telomere length has been investigated as a diagnostic biomarker due to its association with cancer risk and early tumorigenesis.

1. *Associations between telomere length and cancer risk*

   Epidemiological studies have linked telomere length in peripheral blood leukocytes and cancer susceptibility. Both excessively short and long telomeres have been implicated in increased cancer risk:

   - *Short telomeres*: Critically short telomeres can lead to chromosomal instability, promoting oncogenic transformations. This phenomenon is particularly evident in cancers such as lung, bladder, and esophageal cancers [45, 46].

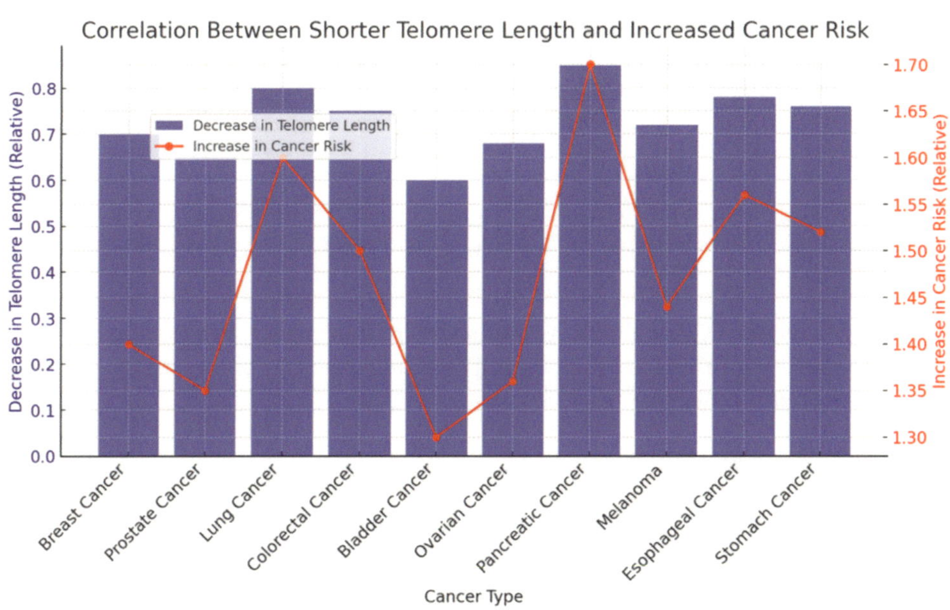

Figure 4.
Telomere length and cancer risk.

- *Long telomeres*: Some studies suggest that unusually long telomeres may provide an extended replicative lifespan to cells, supporting tumor progression. This is observed in cancers like melanoma and certain hematological malignancies [47, 48].

2. *Short telomeres in blood and early detection*

Telomere length in circulating blood cells has been explored as a minimally invasive biomarker for cancer detection. Short telomeres in leukocytes correlate with an elevated risk of developing cancers, including colorectal and gastric cancers [45, 49]. These findings have prompted efforts to integrate telomere measurements into early detection strategies:

- *Population-based screening:* Regular assessment of telomere length could identify individuals at high cancer risk, enabling preventive interventions [50].

- *Liquid biopsy*: Combining telomere length analysis with other circulating biomarkers (e.g., tumor DNA, exosomes) could enhance the sensitivity and specificity of cancer detection [51, 52].

Despite these advances, the diagnostic value of telomere length remains limited by its variability across individuals, tissues, and diseases.

5. Prognostic value of telomere dynamics

Telomere length and dynamics also serve as prognostic indicators, reflecting tumor aggressiveness and patient outcomes (**Table 3**).

1. *Telomere length and tumor aggressiveness*

Studies have shown that cancer patients with extremely short or heterogeneous telomere lengths in tumor tissues often exhibit aggressive disease phenotypes [53–55]:

- *Heterogeneous telomere lengths*: Increased variability in telomere length within tumors indicates genomic instability, a hallmark of high-grade and metastatic cancers.

- *Short telomeres*: Critically short telomeres in tumor cells are associated with enhanced chromosomal fusions and rearrangements, driving tumor progression and metastasis.

2. *Telomere length and patient survival*

Telomere length can provide prognostic insights into overall survival (OS) and progression-free survival (PFS) in various cancers:

- In breast cancer, shorter telomeres are linked to poor prognosis and reduced survival [56, 57].

Cancer Type	Telomere Length/ Telomerase Activity	Prognostic Implications	Study Findings
Glioblastoma	Shortened telomeres, high telomerase activity	Poor overall survival, increased recurrence rate	Short telomeres and high telomerase activity correlate with poor prognosis and shorter survival.
Breast Cancer	Variable telomere length	Short telomeres linked to poor survival and higher recurrence risk	Shorter telomeres are associated with worse outcomes and increased metastasis risk.
Melanoma	High telomerase activity	Poor prognosis, increased metastasis	High telomerase activity is linked to aggressive tumor behavior and lower survival rates.
Non-Small Cell Lung Cancer (NSCLC)	Short telomeres, high telomerase activity	Worse prognosis, higher recurrence rate	Studies show a significant association between short telomeres and poor survival outcomes.
Colorectal Cancer	Shortened telomeres	Poor overall survival, increased metastasis	Short telomeres are indicative of advanced disease stages and poor prognosis.
Prostate Cancer	Variable telomere length	Short telomeres linked to worse survival and increased progression risk	Shorter telomeres are associated with higher Gleason scores and worse outcomes.
Pancreatic Cancer	High telomerase activity	Poor prognosis, higher metastasis rate	High telomerase activity is correlated with advanced tumor stages and reduced survival.
Hepatocellular Carcinoma	Short telomeres, high telomerase activity	Worse survival, higher recurrence	Short telomeres and high telomerase activity are associated with poor prognosis and early recurrence.
Bladder Cancer	Shortened telomeres	Increased recurrence risk, poor survival	Short telomeres are linked to high-grade tumors and reduced survival rates.
Osteosarcoma	High telomerase activity	Poor prognosis, increased metastasis	High telomerase activity is indicative of aggressive tumor behavior and poor clinical outcomes.

Table 3.
Study findings linking telomere length/telomerase activity with patients' prognosis in various cancers.

- Long telomeres may predict favorable outcomes in hematological cancers, as they reflect less genomic instability than short-telomere profiles [58].

- Dynamic changes in telomere length during treatment could also serve as indicators of therapeutic response, with stable telomeres correlating with better outcomes in some cases [59].

The integration of telomere length into prognostic models could improve risk stratification and personalized treatment planning.

6. Technical challenges in telomere biomarker development

Despite its potential, there are significant technical and methodological barriers to implementing telomere-based biomarkers in clinical settings.

1. *Limitations of current telomere measurement techniques*

   Accurately measuring telomere length remains a challenge due to the inherent variability in telomere biology and limitations in available methods:

   - *Quantitative PCR (qPCR):* This widely used technique is cost-effective and has high throughput but suffers from low precision and inter-assay variability [60].

   - *Telomere restriction fragment (TRF) Analysis*: The gold standard TRF assay provides detailed telomere length distributions but is labor-intensive and requires large amounts of DNA [61].

   - *Single telomere length analysis (STELA)*: This technique measures telomere lengths at individual chromosome ends but is limited to certain telomeres and requires specialized expertise [62].

2. *Need for standardization*

   A lack of standardization in telomere measurement protocols and reporting practices hinders the comparison and reproducibility of results across studies. Key areas needing standardization include:

   - *Pre-analytical variables*: Differences in sample collection, storage, and DNA extraction methods can affect telomere length measurements.

   - *Analytical techniques*: Harmonizing methods across laboratories is essential to ensure comparability of results.

   - *Reference ranges*: Establishing population-specific reference ranges for telomere length is critical for meaningful clinical interpretation.

3. *Biological variability*

   Telomere length is influenced by numerous factors, including age, genetic predisposition, environmental exposures, and lifestyle. This variability complicates its use as a standalone biomarker [63].

7. Future directions

To fully realize the potential of telomeres as cancer biomarkers, the following areas warrant further research and development:

DOI: http://dx.doi.org/10.5772/intechopen.1010864

- *Advancing technology*: Developing high-throughput, cost-effective, and precise methods for telomere length measurement, such as next-generation sequencing (NGS)-based approaches.

- *Integrative biomarker panels*: Combining telomere length analysis with other molecular and genetic markers to improve diagnostic and prognostic accuracy.

- *Longitudinal studies*: Establishing the clinical significance of telomere dynamics over time through large-scale prospective studies.

7.1 Therapeutic implications of telomere targeting

Telomere maintenance mechanisms, including telomerase activation and the alternative lengthening of telomeres (ALT), are crucial for tumor cell survival and replication. Targeting these pathways offers a promising therapeutic strategy for a wide range of cancers. While telomerase inhibitors and ALT-targeting therapies have shown potential, their development and application face significant challenges, including resistance mechanisms and tumor heterogeneity. This section explores key advancements, current challenges, and future directions in telomere-targeted cancer therapies.

7.2 Telomerase inhibitors

Telomerase is reactivated in approximately 85–90% of human cancers, making it an attractive therapeutic target [11]. Telomerase inhibitors aim to prevent telomere elongation, leading to telomere shortening and eventual tumor cell death.

1. *Overview of inhibitors and mechanisms of action*

    - *Imetelstat*: A leading telomerase inhibitor, imetelstat is a lipid-conjugated oligonucleotide that directly binds to the RNA component of telomerase (TERC), blocking its enzymatic activity limiting telomere elongation [11, 64, 65].

    - *Small molecule inhibitors*: Compounds like BIBR1532 inhibit the catalytic subunit of telomerase (TERT), disrupting the enzyme's ability to extend telomeres [65, 66].

    - *Telomerase vaccines*: Vaccines targeting TERT elicit immune responses against telomerase-positive cancer cells, potentially enhancing the effectiveness of immunotherapies [67, 68].

2. *Clinical trial outcomes and ongoing developments*

    - *Imetelstat*: Early clinical trials demonstrated its efficacy in hematological malignancies such as myelofibrosis, which reduced telomerase activity and prolonged overall survival in certain patients [69–71]. However, challenges like dose-limiting toxicities, including cytopenias, remain a concern.

    - Combination therapies involving imetelstat and other agents are under investigation to improve therapeutic efficacy and reduce resistance [72, 73].

Therapy	Indication	Phase	Status	Details
THIO (6-thio-2'-deoxyguanosine)	Advanced Non-Small Cell Lung Cancer (NSCLC)	Phase 2	Ongoing	MAIA Biotechnology is conducting the THIO-101 trial to evaluate THIO in combination with the anti-PD-1 antibody cemiplimab. The U.S. FDA cleared the IND application in October 2023. BioSpace Updated results were presented at the 2024 ASCO Annual Meeting. ASCOPubs
EXG-34217	Dyskeratosis Congenita	Phase 1/2	Ongoing	Elixirgen Therapeutics is assessing EXG-34217, which involves ex vivo telomere elongation of autologous CD34+ hematopoietic stem cells using EXG-001. Positive interim data were presented at the 64th ASH Annual Meeting in December 2022. Elixirgen Therapeutics

Table 4.
Ongoing clinical trials for telomere-targeting therapies as of January 17, 2025.

- Trials in solid tumors have been less successful, potentially due to the slower kinetics of telomere shortening in these cancers compared to rapidly dividing hematological malignancies (**Table 4**) [74, 75].

7.3 Targeting ALT pathways

The ALT mechanism, employed by about 10–15% of cancers, particularly in sarcomas and gliomas, presents a distinct therapeutic opportunity [37].

1. *Therapeutic opportunities in ALT-positive cancers*

- ALT is characterized by telomere length heterogeneity and the presence of ALT-associated promyelocytic leukemia (PML) bodies.

- *ATR inhibitors*: ATR (ataxia-telangiectasia and Rad3-related) kinase is essential for the survival of ALT-positive cells. Inhibitors like VE-821 and VE-822 disrupt ALT pathways by impairing DNA damage response signaling, selectively killing ALT-positive tumor cells [76, 77].

- Other strategies include targeting proteins involved in homologous recombination, such as RAD51 or FANCM, which are critical for ALT-mediated telomere elongation [24, 31].

2. *Preclinical studies and challenges in drug development*

- Challenges include the lack of reliable biomarkers for ALT status and potential off-target effects, given the role of ATR in normal DNA damage repair.

- The redundancy of ALT-related pathways also complicates the development of highly specific therapies.

7.4 Combination therapies

Combining telomere-targeting agents with existing cancer therapies has shown promise in overcoming some of their limitations.

1. *Synergistic potential with chemotherapy or immunotherapy*

    - Telomerase inhibitors enhance the efficacy of chemotherapy by exacerbating DNA damage in tumor cells. For example, combining imetelstat with paclitaxel or carboplatin has shown additive effects in preclinical models [66, 78].

    - Vaccines targeting TERT, when combined with immune checkpoint inhibitors (e.g., anti-PD-1/PD-L1 antibodies), boost antitumor immune responses, particularly in immunologically "cold" tumors [67, 79].

2. Enhancing sensitivity to DNA-damaging agents

    - Telomere-targeting agents increase tumor cell sensitivity to radiation and DNA-damaging agents, such as cisplatin, by compromising telomere integrity.

    - ALT-positive cancers, in particular, are highly dependent on robust DNA damage responses. Inhibiting ALT mechanisms further sensitizes these tumors to genotoxic therapies [80, 81].

7.5 Resistance mechanisms

Despite their potential, telomere-targeting therapies face significant hurdles due to the adaptability of cancer cells.

1. *Adaptive responses in tumors*

    - Tumors often develop resistance to telomerase inhibitors through upregulation of alternative telomere maintenance pathways, including ALT activation.

    - Increased DNA repair efficiency and activation of bypass mechanisms allow cancer cells to evade the effects of telomere-targeted therapies [24, 82].

2. *Strategies to overcome resistance*

    - *Combination approaches*: Using telomerase inhibitors alongside ALT pathway inhibitors or DNA repair inhibitors (e.g., PARP inhibitors) can mitigate resistance [83–85].

    - *Targeting telomere-associated proteins:* Disrupting shelterin components, such as TRF1 and TRF2, could compromise telomere stability, regardless of the maintenance mechanism [84, 86].

    - *Adaptive dosing regimens*: Alternating or intermittent dosing schedules may prevent the development of resistance by minimizing selective pressures [87, 88].

The development of telomere-targeting therapies is a rapidly evolving field, with several areas requiring further research and optimization:

- *Biomarker development*: Reliable biomarkers for telomerase activity or ALT status are needed to identify patients who are most likely to benefit from these therapies.

- *Combination strategies*: Systematic evaluation of combination therapies involving telomere-targeting agents and standard cancer treatments is crucial.

- *Personalized approaches*: Tailoring telomere-targeting therapies based on individual tumor biology and telomere maintenance mechanisms will enhance efficacy.

7.6 Emerging insights in telomere biology and cancer

Recent advances in the study of telomeres have revealed noncanonical roles of telomerase, beyond its traditional function in telomere elongation, and highlighted the contribution of telomere-associated proteins in cancer progression and resistance to therapy. Additionally, growing evidence underscores the importance of epigenetic and transcriptomic regulation in maintaining telomere integrity and influencing tumorigenesis. These emerging insights are reshaping our understanding of telomere biology and opening new avenues for targeted cancer therapies.

7.7 Noncanonical roles of telomerase

1. *Telomerase in stem cell maintenance*

   Telomerase has long been recognized for its role in maintaining telomere length in somatic cells, but its involvement in stem cell biology is gaining increased attention. In stem cells, telomerase activity is essential for maintaining self-renewal and preventing premature senescence. Recent studies have shown that telomerase elongates telomeres and regulates stem cell homeostasis by interacting with key signaling pathways like Wnt, Notch, and TGF-β [89–91]. These pathways are critical for the balance between stem cell renewal and differentiation, and telomerase is increasingly viewed as a regulator of cellular pluripotency and tissue regeneration [92]. Furthermore, telomerase may play a role in maintaining the balance between stem cell quiescence and activation, suggesting its significance in cancer stem cell populations [93, 94].

2. *Telomerase in mitochondrial function*

   Telomerase has been implicated in maintaining mitochondrial function, extending beyond its role in the nucleus. Studies have demonstrated that telomerase reverse transcriptase (TERT), the catalytic subunit of telomerase, localizes to mitochondria and contributes to mitochondrial stability and bioenergetics [95]. TERT has been shown to regulate mitochondrial DNA integrity and protect against oxidative stress, thereby supporting cellular energy metabolism. This mitochondrial function of TERT is particularly relevant in cancer cells, where mitochondrial dysfunction is common. By stabilizing mitochondrial DNA and preventing cellular stress, telomerase supports the survival and proliferation of cancer cells in hypoxic environments, thus contributing to tumorigenesis [96].

3. *Telomerase in signaling pathways*

Beyond its involvement in stem cells and mitochondria, telomerase has been shown to interact with several cellular signaling pathways, including the PI3K/AKT, MAPK, and NF-κB pathways. These interactions suggest that telomerase plays a role in regulating cell survival, apoptosis, and proliferation, all of which are crucial for cancer progression [97, 98]. In particular, the ability of telomerase to modulate these pathways in response to DNA damage or oxidative stress contributes to its role in cancer cell survival and resistance to chemotherapy. Telomerase may also promote epithelial-to-mesenchymal transition (EMT), a key event in metastasis, by regulating transcription factors involved in cell migration and invasion [28, 99].

8. Telomere-associated proteins

1. *Roles of shelterin components in cancer progression*

The shelterin complex, composed of six proteins (TRF1, TRF2, RAP1, POT1, TIN2, and TPP1), plays a critical role in protecting telomeres from degradation and inappropriate DNA repair. In cancer, dysregulation of shelterin components is frequently observed, leading to telomere dysfunction, genomic instability, and uncontrolled cell proliferation. For example, overexpression of TRF1 or TRF2 is commonly observed in various cancers and is associated with telomere protection and tumor progression [86, 100]. Conversely, loss of shelterin proteins such as POT1 has been linked to the activation of the DNA damage response, resulting in telomere fusion and chromosomal instability, a hallmark of cancer progression [101, 102].

Shelterin proteins also contribute to the resistance of cancer cells to therapies that induce DNA damage. TRF1, for instance, has been shown to inhibit the activation of the DNA damage response at telomeres, thus preventing the triggering of cell death pathways following radiation or chemotherapy. This ability to evade apoptotic signals is a key mechanism by which cancer cells resist conventional therapies [102–104].

2. *Epigenetic regulation of shelterin complex components*

Recent studies have uncovered epigenetic mechanisms that regulate the expression and activity of shelterin components. For instance, histone modifications and DNA methylation are involved in the regulation of TRF1 and TRF2 expression, thereby influencing telomere length maintenance and cancer progression [105]. The upregulation of these shelterin components *via* epigenetic changes can contribute to the evasion of senescence and promote tumorigenesis by stabilizing telomere integrity in cancer cells [106, 107].

3. *Therapeutic implications of targeting shelterin proteins*

Given their pivotal role in maintaining telomere integrity, shelterin proteins are attractive targets for therapeutic intervention. Inhibition of shelterin proteins such as TRF1 or POT1 may render cancer cells more susceptible to DNA damage and cell death, providing a potential strategy for cancer therapy. However, the development of therapies targeting shelterin components must account for the potential toxicity to normal cells, where these proteins play essential roles in maintaining genomic stability [108, 109].

9. Epigenetic and transcriptomic regulation of telomeres

1. *Chromatin remodeling and telomere dynamics*

Chromatin remodeling is a crucial process for regulating telomere maintenance and function. Modifications to histones and chromatin structure can influence the accessibility of telomeres to telomerase and other repair machinery. For instance, histone acetylation at telomeres has been shown to promote telomere elongation by facilitating telomerase binding [110, 111]. On the other hand, histone deacetylation can lead to telomere attrition and genomic instability, providing a link between chromatin remodeling and cancer development. Furthermore, the interaction between telomeres and various chromatin remodeling complexes, such as the SWI/SNF complex, has been implicated in telomere regulation during cell division [112, 113].

2. *Noncoding RNAs and telomere regulation*

The role of noncoding RNAs (ncRNAs) in telomere biology is an emerging area of research. Telomere-specific ncRNAs, including TERRA (telomeric repeat-containing RNA), have been shown to regulate telomere stability by modulating the chromatin structure at telomeres. TERRA is involved in the formation of telomere heterochromatin and the regulation of telomerase activity. Dysregulation of TERRA and other ncRNAs can lead to telomere dysfunction and contribute to the development of cancer by promoting genomic instability [114, 115]. Moreover, long noncoding RNAs (lncRNAs) such as HOTAIR and MALAT1, which are involved in the regulation of gene expression, have been found to interact with telomerase and contribute to telomere maintenance in cancer cells [116–118].

3. *Telomere and epigenetic reprogramming in cancer*

Epigenetic reprogramming, which alters the expression of genes involved in telomere maintenance, is a hallmark of cancer. The dysregulation of epigenetic processes, such as DNA methylation and histone modification, can result in the silencing or activation of genes that regulate telomerase and shelterin proteins, leading to altered telomere dynamics in cancer cells [110, 119]. For example, the silencing of the tumor suppressor gene p16INK4a, which regulates telomere shortening, is often observed in various cancers and is associated with telomere elongation and enhanced tumorigenicity [120, 121].

Emerging insights into telomere biology reveal that telomerase and telomere-associated proteins play critical roles beyond telomere elongation, including regulating stem cell function, mitochondrial health, and cellular signaling pathways. The epigenetic regulation of telomere dynamics further complicates the process, contributing to cancer progression and therapeutic resistance. These discoveries provide novel therapeutic opportunities, such as targeting telomere-associated proteins and epigenetic pathways, which could lead to more effective cancer treatments. However, challenges remain in translating these findings into clinical practice, particularly in terms of specificity and minimizing off-target effects. Further research into the noncanonical roles of telomerase and the regulatory mechanisms that influence telomere function will be critical for the development of next-generation cancer therapies.

10. Future directions in telomere research

Telomeres, as critical elements of cellular stability and aging, continue to garner attention as potential therapeutic targets in cancer. As our understanding of telomere biology deepens, emerging technologies and strategies are opening new avenues for cancer treatment. The application of advanced techniques such as single-cell telomere length analysis and CRISPR-based tools is poised to revolutionize our approach to understanding telomere dynamics in cancer. Furthermore, the integration of telomere-targeting therapies into personalized medicine and the exploration of telomere-based interventions hold significant promise for improving cancer treatment outcomes. However, several unanswered questions still need to be addressed to fully harness the potential of telomere-targeting therapies.

11. Technological advancements in telomere research

1. *Single-cell telomere length analysis*

   Traditional methods of telomere length measurement, such as quantitative PCR (qPCR) or Southern blotting, have provided valuable insights into telomere dynamics. However, they typically measure average telomere lengths across populations of cells, which masks the heterogeneity present in individual cells. Recent advancements in single-cell telomere length analysis have enabled the study of telomere dynamics at the level of individual cancer cells. Techniques such as single-cell telomere sequencing and single-cell qPCR now allow for the assessment of telomere length heterogeneity within tumor populations [61, 122].
   This technological leap has profound implications for understanding tumor heterogeneity, which is a key factor in cancer progression and therapeutic resistance. Tumors often contain subpopulations of cells with varying telomere lengths, which may contribute to differential responses to therapies. For instance, cells with short telomeres may undergo senescence or die in response to treatment, while those with longer telomeres may continue to proliferate. By analyzing telomere length at the single-cell level, researchers can gain insights into the clonal evolution of tumors, identify biomarkers for drug resistance, and predict relapse. This approach holds promise for improving patient stratification in clinical settings [123, 124].

2. *CRISPR-based tools for studying telomere dynamics*

   The CRISPR/Cas9 genome-editing system has emerged as a powerful tool for investigating telomere biology. By selectively targeting genes involved in telomere maintenance, such as those encoding shelterin proteins or telomerase

subunits, researchers can modulate telomere length and study its effects on cellular processes, including cancer progression. CRISPR-based systems have enabled the precise manipulation of telomere length and telomerase activity, facilitating the study of the molecular pathways that regulate telomere dynamics [125–127].
Additionally, CRISPR/Cas9 can be used to engineer cancer models with specific telomere-related mutations, such as mutations in the TERT promoter or mutations that activate the ALT pathway. These models provide valuable insights into how telomere alterations contribute to oncogenesis and resistance to conventional therapies. Furthermore, CRISPR-based tools are also being developed to specifically target and disrupt telomere-maintenance pathways in cancer cells, providing a potential therapeutic approach to eliminate cancer cells with dysfunctional telomeres selectively [127, 128].

12. Personalized medicine and telomere-targeting therapies

1. *Integrating telomere-targeting strategies into tailored treatment regimens*

As our understanding of the role of telomere maintenance in cancer grows, the possibility of integrating telomere-targeting strategies into personalized medicine is becoming increasingly feasible. Personalized medicine aims to tailor treatments based on the individual genetic and molecular characteristics of a patient's tumor. Telomere biology plays a crucial role in determining the aggressiveness of cancer and its response to therapy. By assessing the telomere status of tumors, clinicians can gain insights into the most appropriate treatment strategies.
For example, in cancers that rely on telomerase for telomere maintenance, telomerase inhibitors (such as imetelstat) may be particularly effective. Alternatively, tumors with ALT activity may benefit from therapies targeting the alternative lengthening of the telomeres (ALT) pathway. Combining telomere-targeting therapies with other treatment modalities, such as chemotherapy or immunotherapy, could enhance the effectiveness of treatment regimens and reduce the likelihood of relapse. Integrating telomere length assessment and telomerase activity measurements into routine clinical practice may enable clinicians to select the most effective personalized treatment strategies [128–130].

2. *Biomarkers for telomere-based interventions*

As telomere biology becomes more integrated into clinical practice, the identification of reliable biomarkers for telomere-targeting therapies will be essential. Telomere length and telomere-associated proteins, such as shelterin components, could serve as biomarkers to predict the effectiveness of telomere-targeted therapies. Moreover, the detection of telomerase activity or the presence of ALT mechanisms in tumors could provide valuable insights into treatment response. Combining these biomarkers with other molecular features of tumors, such as genetic mutations and epigenetic modifications, will help create more accurate prognostic tools and improve patient outcomes [23, 31, 37].

DOI: http://dx.doi.org/10.5772/intechopen.1010864

13. Unanswered questions in telomere research

1. *Role of telomere dynamics in treatment resistance and relapse*

   One of the major challenges in cancer therapy is the development of treatment resistance and relapse. While telomere maintenance mechanisms are implicated in these processes, the exact role of telomere dynamics in resistance remains poorly understood. In particular, it is unclear how telomere attrition or telomere elongation contributes to resistance to conventional therapies, such as chemotherapy and radiation, as well as newer immunotherapies. Some studies suggest that telomere dysfunction may trigger a DNA damage response that promotes tumor survival, while others indicate that long telomeres may allow cancer cells to bypass replicative senescence and continue proliferating despite treatment. Further research is needed to better understand the relationship between telomere dynamics and therapeutic resistance. Investigating how telomere alterations influence DNA repair mechanisms, cell cycle progression, and apoptotic pathways in cancer cells will be critical for developing strategies to overcome resistance and prevent relapse [37, 131, 132].

2. *Telomere-based interventions in combination with emerging cancer therapies*

   The combination of telomere-targeting therapies with emerging cancer therapies, such as immune checkpoint inhibitors and targeted therapies, holds great potential. However, the safety and efficacy of such combination approaches are still unclear. Telomere-targeting therapies could potentially sensitize cancer cells to immune responses by inducing telomere dysfunction and triggering cell death pathways. On the other hand, telomere-targeting drugs may also have off-target effects on normal cells, raising concerns about toxicity and side effects. Understanding the optimal combination strategies and determining the best patient populations for these treatments will require extensive clinical trials and preclinical studies. Additionally, the timing of telomere-targeted therapies in combination with other treatments needs to be explored to maximize therapeutic efficacy while minimizing potential resistance mechanisms [133–135].
   The future of telomere research holds tremendous potential for improving cancer therapies and personalizing treatment regimens. Advances in technologies such as single-cell telomere analysis and CRISPR-based tools are providing new insights into tumor heterogeneity and telomere dynamics, which could help refine cancer diagnosis and treatment. Integrating telomere-targeting strategies into personalized medicine offers the promise of more effective and tailored therapies. However, several unanswered questions remain, particularly regarding the role of telomere dynamics in treatment resistance and relapse. Continued research is essential to understand the complexities of telomere maintenance in cancer and to develop effective therapies that target telomeres and related pathways.

14. Conclusion

In this chapter, we explored the crucial role of telomere biology in cancer, examining the complex mechanisms of telomere maintenance, their implications in cancer progression, and the potential for targeting these mechanisms in therapeutic settings.

The dynamic processes governing telomere length, such as the activation of telomerase and alternative lengthening of telomeres (ALT), not only are essential for maintaining chromosomal stability but also serve as key drivers of tumorigenesis and resistance to conventional treatments. Telomeres, once considered merely structural elements at the ends of chromosomes, have now emerged as critical components in cancer progression, with their dynamics influencing tumor growth, metastasis, and treatment response.

14.1 Telomere maintenance mechanisms as critical drivers of cancer progression

Telomere maintenance mechanisms, primarily through the activation of telomerase or the ALT pathway, provide a survival advantage to cancer cells by bypassing replicative senescence and enabling continuous proliferation. This dysregulation of telomere biology is often observed in a variety of malignancies, making it a hallmark of cancer. The aberrant reactivation of telomerase is a common feature in over 85% of human cancers and is strongly associated with increased malignancy and poor prognosis [46, 136, 137]. Similarly, cancers utilizing the ALT mechanism are characterized by increased genomic instability, further driving tumor evolution and resistance to therapies [138–140]. These findings underscore the significance of telomere maintenance pathways as not only a consequence of cancer but also active participants in the malignancy's development and progression.

14.2 The potential of telomeres as biomarkers and therapeutic targets

Telomeres, due to their role in cellular aging and genomic stability, hold great promise as biomarkers for cancer diagnosis, prognosis, and therapy selection. Telomere length has been associated with both cancer risk and prognosis, with shorter telomeres in peripheral blood correlated with increased susceptibility to cancer and shorter survival in patients with certain cancers [141–144]. As telomere shortening is a ubiquitous feature of aging and cancer, measuring telomere dynamics may offer an early detection tool for malignancies, particularly in cancers known to be associated with significant telomere attrition, such as lung, liver, and colorectal cancers [145, 146]. Moreover, assessing telomere-related biomarkers such as telomerase activity or shelterin complex expression may provide crucial insights into the biological behavior of tumors and their response to treatment [45, 47].

Therapeutically, telomere-targeting strategies, including telomerase inhibitors like imetelstat and approaches targeting the ALT pathway, hold promise as potential cancer treatments. By disrupting these pathways, which cancer cells rely on for unlimited proliferation, researchers aim to selectively target tumor cells while sparing normal somatic cells. Combination therapies involving telomere-targeting agents and traditional chemotherapies, or newer immunotherapies, may also improve treatment efficacy by sensitizing cancer cells to DNA-damaging agents and immune responses [64, 147, 148]. These therapeutic strategies are still in the experimental phase, and while preclinical studies show promising results, clinical trials are required to assess their true therapeutic potential and safety in patients.

14.3 Call for interdisciplinary research to translate insights into clinical practice

The future of telomere-targeting therapies hinges on our ability to translate fundamental research into clinical practice. This will require interdisciplinary collaboration among molecular biologists, oncologists, bioinformaticians, and clinical researchers. While significant advances have been made in understanding telomere biology, challenges remain in accurately measuring telomere length, understanding the full spectrum of telomere-associated proteins, and elucidating the mechanisms that govern telomere maintenance across various cancers. As telomere-targeting therapies move from preclinical studies to clinical trials, it is essential to standardize methodologies for telomere measurement and biomarker identification to ensure reliable, reproducible results across different patient populations [149, 150].

Furthermore, the development of individualized treatment regimens, taking into account the telomere status of each patient's tumor, will be crucial for maximizing the effectiveness of telomere-targeted therapies. This requires the integration of telomere-related biomarkers into routine cancer diagnostics and prognostics, which will likely lead to more personalized and effective therapeutic approaches. Ultimately, understanding how telomere maintenance contributes to treatment resistance and relapse will be key to improving long-term outcomes for cancer patients [151].

Telomeres and their associated pathways are central to cancer biology and hold great promise for both diagnostics and treatment. However, translating these insights into clinical therapies requires continued research, collaboration, and innovation. As our understanding of telomere dynamics grows, so too will our ability to harness this knowledge to improve cancer diagnosis, prognosis, and therapeutic strategies, providing new hope for patients with cancer.

Acknowledgements

The author wishes to express gratitude to all researchers and scholars whose work has contributed to understanding this topic. This chapter was conducted independently, and while no direct collaborations or funding were involved, the insights gained from the broader academic community have been invaluable. The author also appreciates family and friends for their encouragement and support throughout the writing process.

The author acknowledges the use of ChatGPT for language polishing of the chapter.

Funding

This research received no specific grant from any funding agencies.

Conflict of interests

The author declares no conflicts of interest related to this book chapter. This work was conducted independently, with no financial, personal, or professional relationships that could have influenced its content, findings, or interpretations. Additionally, no external funding or institutional obligations impacted the integrity or objectivity of this chapter.

Abbreviations

AKT	protein kinase B
ALT	alternative lengthening of telomeres
ATM	ataxia telangiectasia mutated
ATR	ATM and Rad3-related
BBA	biochimica et biophysica acta
BIR	break-induced replication
BMC	BioMed central
CRISPR	clustered regularly interspaced short palindromic repeats
CST	CTC1-STN1-TEN1 complex
DDR	DNA damage response
DNA	deoxyribonucleic acid
EMBO	European molecular biology organization
EMT	epithelial-mesenchymal transition
ETS/TCF	E26 transformation-specific/ternary complex factor
FANCM	fanconi anemia complementation group M
FEBS	federation of European biochemical societies
FISH	fluorescent in situ hybridization
HBV	hepatitis B virus
HCV	hepatitis C virus
HER2	human epidermal growth factor receptor 2
HNSCC	head and neck squamous cell carcinoma
HPV	human papillomavirus
IDH	isocitrate dehydrogenase
JNCI	journal of the National Cancer Institute
MAPK	mitogen-activated protein kinase
MYC	myelocytomatosis
NF	nuclear factor
NGS	next-generation sequencing
NSCLC	non-small cell lung cancer
OS	overall survival
PARP	poly (ADP-ribose) polymerase
PCR	polymerase chain reaction
PD-1/PD	programmed death protein 1/programmed death
PFS	progression-free survival
P13K	phosphoinositide 3-kinase
PML	promyelocytic leukemia
POT1	protection of telomeres 1
RAP1	repressor/activator protein 1
RAS	rat sarcoma virus
qPCR	quantitative polymerase chain reaction
RB	retinoblastoma
RNA	ribonucleic acid
SASP	senescence-associated secretory phenotype
SCE	sister chromatid exchange
STELA	single telomere length analysis
SWI/SNF	switch (mating-type)/**s**ucrose non-fermenting (**SWI**tch)
TERC	telomerase RNA component
TERRA	telomeric repeat-containing RNA

TERT	telomerase reverse transcriptase
TGF	transforming growth factor
TICCA	transient, immediate, complete, and combinatory attack
TIN2	TERT1-interacting nuclear factor 2
TPP1	tripeptidyl peptidase 1
TRF1	telomeric repeat- binding factor 1
TRF2	telomeric repeat- binding factor 2
TTAGGG	thymine–thymine–adenine–guanine–guanine–guanine
VE-821	vertex experimental compound 821
VE-822	vertex experimental compound 822
VE-970	vertex experimental compound 970
WNT	wingless-related integration site

Author details

Nosipho Treasure Mdluli[1] and Yun-Chen Chang[2,3]*

1 Graduate Institute of Aging, China Medical University, Taichung, Taiwan

2 Graduate Institute of Nursing, China Medical University, Taichung, Taiwan

3 Nursing Department, China Medical University Hospital, Taichung, Taiwan

*Address all correspondence to: lisacow@mail.cmu.edu.tw; lisacow@hotmail.com

References

[1] Blackburn EH. Structure and function of telomeres. Nature. 1991;**350**(6319):569-573

[2] Shay JW, Wright WE. Telomeres and telomerase in normal and cancer stem cells. FEBS Letters. 2010;**584**(17):3819-3825

[3] Hanahan D, Weinberg RA. Hallmarks of cancer: The next generation. Cell. 2011;**144**(5):646-674

[4] Shay JW, Bacchetti S. A survey of telomerase activity in human cancer. European Journal of Cancer. 1997;**33**(5):787-791

[5] Cesare AJ, Reddel RR. Alternative lengthening of telomeres: Models, mechanisms and implications. Nature Reviews Genetics. 2010;**11**(5):319-330

[6] Flynn RL et al. Alternative lengthening of telomeres renders cancer cells hypersensitive to ATR inhibitors. Science. 2015;**347**(6219):273-277

[7] Weischer M et al. Short telomere length, cancer survival, and cancer risk in 47102 individuals. Journal of the National Cancer Institute. 2013;**105**(7):459-468

[8] Meena JK et al. Telomerase abrogates aneuploidy-induced telomere replication stress, senescence and cell depletion. The EMBO Journal. 2015;**34**(10):1371-1384

[9] Shay JW. Role of telomeres and telomerase in aging and cancer. Cancer Discovery. 2016;**6**(6):584-593

[10] de Jesus BB, Blasco MA. Telomerase at the intersection of cancer and aging. Trends in Genetics. 2013;**29**(9):513-520

[11] Jafri MA et al. Roles of telomeres and telomerase in cancer, and advances in telomerase-targeted therapies. Genome Medicine. 2016;**8**:1-18

[12] Conomos D et al. Variant repeats are interspersed throughout the telomeres and recruit nuclear receptors in ALT cells. Journal of Cell Biology. 2012;**199**(6):893-906

[13] De Lange T. Shelterin: The protein complex that shapes and safeguards human telomeres. Genes and Development. 2005;**19**(18):2100-2110

[14] Smogorzewska A, de Lange T. Regulation of telomerase by telomeric proteins. Annual Review of Biochemistry. 2004;**73**(1):177-208

[15] Griffith JD et al. Mammalian telomeres end in a large duplex loop. Cell. 1999;**97**(4):503-514

[16] Artandi SE, DePinho RA. Telomeres and telomerase in cancer. Carcinogenesis. 2010;**31**(1):9-18

[17] Fagagna FDAD et al. A DNA damage checkpoint response in telomere-initiated senescence. Nature. 2003;**426**(6963):194-198

[18] Rodier F, Campisi J. Four faces of cellular senescence. Journal of Cell Biology. 2011;**192**(4):547-556

[19] Coppé J-P et al. Senescence-associated secretory phenotypes reveal cell-nonautonomous functions of oncogenic RAS and the p53 tumor suppressor. PLoS Biology. 2008;**6**(12):e301

[20] Huang FW et al. Highly recurrent TERT promoter mutations in

human melanoma. Science. 2013;**339**(6122):957-959

[21] Horn S et al. TERT promoter mutations in familial and sporadic melanoma. Science. 2013;**339**(6122):959-961

[22] Zhang Y et al. E2F1: A potential negative regulator of hTERT transcription in normal cells upon activation of oncogenic c-Myc. Medical Science Monitor: International Medical Journal of Experimental and Clinical Research. 2012;**18**(1):RA12

[23] Sohn EJ et al. The molecular mechanisms and therapeutic prospects of alternative lengthening of telomeres (ALT). Cancers. 2023;**15**(7):1945

[24] Gao J, Pickett HA. Targeting telomeres: Advances in telomere maintenance mechanism-specific cancer therapies. Nature Reviews Cancer. 2022;**22**(9):515-532

[25] Olympios N et al. TERT promoter alterations in glioblastoma: A systematic review. Cancers. 2021;**13**(5):1147

[26] Yuan X, Larsson C, Xu D. Mechanisms underlying the activation of TERT transcription and telomerase activity in human cancer: Old actors and new players. Oncogene. 2019;**38**(34):6172-6183

[27] Donate LE, Blasco MA. Telomeres in cancer and ageing. Philosophical Transactions of the Royal Society B: Biological Sciences. 2011;**366**(1561):76-84

[28] Hao Y, Baker D, Ten Dijke P. TGF-β-mediated epithelial-mesenchymal transition and cancer metastasis. International Journal of Molecular Sciences. 2019;**20**(11):2767

[29] Ivancich M et al. Treating cancer by targeting telomeres and telomerase. Antioxidants. 2017;**6**(1):15

[30] Fernandes SG et al. Role of telomeres and telomeric proteins in human malignancies and their therapeutic potential. Cancers. 2020;**12**(7):1901

[31] Sun H et al. Potential clinical treatment prospects behind the molecular mechanism of alternative lengthening of telomeres (ALT). Journal of Cancer. 2023;**14**(3):417

[32] Stögbauer L et al. Telomerase activity, TERT expression, hTERT promoter alterations, and alternative lengthening of the telomeres (ALT) in meningiomas–a systematic review. Neurosurgical Review. 2020;**43**:903-910

[33] Royle NJ et al. The Role of Recombination in Telomere Length Maintenance. Portland Press Ltd.; 2009

[34] Anand RP, Lovett ST, Haber JE. Break-induced DNA replication. Cold Spring Harbor Perspectives in Biology. 2013;**5**(12):a010397

[35] Abid HZ et al. Single-molecule analysis of subtelomeres and telomeres in alternative lengthening of telomeres (ALT) cells. BMC Genomics. 2020;**21**:1-17

[36] Chung I et al. PML body meets telomere: The beginning of an ALTernate ending? Nucleus. 2012;**3**(3):263-275

[37] MacKenzie D Jr et al. ALT positivity in human cancers: Prevalence and clinical insights. Cancers. 2021;**13**(10):2384

[38] Claude E et al. Detection of alternative lengthening of telomeres mechanism on tumor sections. Molecular Biomedicine. 2021;**2**:1-16

[39] Henson JD et al. The C-circle assay for alternative-lengthening-of-telomeres activity. Methods. 2017;**114**:74-84

[40] Jiang W-Q et al. Protocol III: Detection of alternative lengthening of telomeres. Telomeres and Telomerase in Cancer. 2009:351-364

[41] Yeager TR et al. Telomerase-negative immortalized human cells contain a novel type of promyelocytic leukemia (PML) body. Cancer Research. 1999;**59**(17):4175-4179

[42] Forood AMK. Mechanisms of Telomere Dysfunction in Cancer from Genomic Instability to Therapy: A. 2024

[43] Di Marco T et al. Non-oncogene dependencies: Novel opportunities for cancer therapy. Biochemical Pharmacology. 2024:116254

[44] Harvey AJ. The Role of DNA Repair and Regulatory Proteins in the Maintenance of Human Telomeres and their Control of Cellular Immortalization. University of Minnesota; 2017

[45] Loukopoulou C et al. Telomere length and telomerase activity as potential biomarkers for gastrointestinal cancer. Cancers. 2024;**16**(19):3370

[46] Trybek T et al. Telomeres and telomerase in oncogenesis. Oncology Letters. 2020;**20**(2):1015-1027

[47] Srinivas N, Rachakonda S, Kumar R. Telomeres and telomere length: A general overview. Cancers. 2020;**12**(3):558

[48] McNally EJ, Luncsford PJ, Armanios M. Long telomeres and cancer risk: The price of cellular immortality. The Journal of Clinical Investigation. 2019;**129**(9):3474-3481

[49] Heaphy CM, Meeker AK. The potential utility of telomere-related markers for cancer diagnosis. Journal of Cellular and Molecular Medicine. 2011;**15**(6):1227-1238

[50] Naing C et al. Association between telomere length and the risk of colorectal cancer: A meta-analysis of observational studies. BMC Cancer. 2017;**17**:1-7

[51] Dabral P et al. Tumor-derived extracellular vesicles as liquid biopsy for diagnosis and prognosis of solid tumors: Their clinical utility and reliability as tumor biomarkers. Cancers. 2024;**16**(13):2462

[52] Wu J et al. Tumor circulome in the liquid biopsies for cancer diagnosis and prognosis. Theranostics. 2020;**10**(10):4544

[53] Bisoffi M, Heaphy CM, Griffith JK. Telomeres: Prognostic markers for solid tumors. International Journal of Cancer. 2006;**119**(10):2255-2260

[54] Tsatsakis A et al. Role of telomere length in human carcinogenesis. International Journal of Oncology. 2023;**63**(1):1-24

[55] Diaz-Cano SJ. Tumor heterogeneity: Mechanisms and bases for a reliable application of molecular marker design. International Journal of Molecular Sciences. 2012;**13**(2):1951-2011

[56] Shen J et al. Genetic polymorphisms in telomere pathway genes, telomere length, and breast cancer survival. Breast Cancer Research and Treatment. 2012;**134**:393-400

[57] Ennour-Idrissi K, Maunsell E, Diorio C. Telomere length and breast cancer prognosis: A systematic review. Cancer Epidemiology, Biomarkers & Prevention. 2017;**26**(1):3-10

[58] Vasko T et al. Telomeres and telomerase in hematopoietic dysfunction: Prognostic implications and pharmacological interventions. International Journal of Molecular Sciences. 2017;**18**(11):2267

[59] Gallicchio L et al. The effect of cancer treatments on telomere length: A systematic review of the literature. JNCI: Journal of the National Cancer Institute. 2018;**110**(10):1048-1058

[60] Morinha F, Magalhães P, Blanco G. Standard guidelines for the publication of telomere qPCR results in evolutionary ecology. Molecular Ecology Resources. 2020;**20**(3):635-648

[61] Yu HJ, Byun YH, Park C-K. Techniques for assessing telomere length: A methodological review. Computational and Structural Biotechnology Journal. 2024

[62] Aubert G, Hills M, Lansdorp PM. Telomere length measurement—Caveats and a critical assessment of the available technologies and tools. Mutation Research/Fundamental and Molecular Mechanisms of Mutagenesis. 2012;**730**(1-2):59-67

[63] Fasching CL. Telomere length measurement as a clinical biomarker of aging and disease. Critical Reviews in Clinical Laboratory Sciences. 2018;**55**(7):443-465

[64] Xu Y, Goldkorn A. Telomere and telomerase therapeutics in cancer. Genes. 2016;7(6):22

[65] Romaniuk A et al. Telomerase modulation in therapeutic approach. Current Pharmaceutical Design. 2014;**20**(41):6438-6451

[66] Vishwakarma K, Dey R, Bhatt H. Telomerase: A prominent oncological target for development of chemotherapeutic agents. European Journal of Medicinal Chemistry. 2023;**249**:115121

[67] Ellingsen EB. Telomerase Based Therapeutic Vaccination and Checkpoint Inhibition: Characterization of the Induced Immune Response and Impact on the Tumor Microenvironment. 2023

[68] Vahidi S, Touchaei AZ. Telomerase-based vaccines: A promising frontier in cancer immunotherapy. Cancer Cell International. 2024;**24**:421

[69] Ziedan M, Salah D, Ayman M. Therapeutic impact of telomerase inhibitor Imetelstat: A literature review. International Journal of Advanced Health Science and Technology. 2023;**3**(2):111-117

[70] Mascarenhas J et al. Randomized, single-blind, multicenter phase II study of two doses of imetelstat in relapsed or refractory myelofibrosis. Journal of Clinical Oncology. 2021;**39**(26):2881-2892

[71] Waksal JA et al. Telomerase-targeted therapies in myeloid malignancies. Blood Advances. 2023;**7**(16):4302-4314

[72] Koziel JE, Herbert B-S. The telomerase inhibitor imetelstat alone, and in combination with trastuzumab, decreases the cancer stem cell population and self-renewal of HER2+ breast cancer cells. Breast Cancer Research and Treatment. 2015;**149**:607-618

[73] Fischer-Mertens J et al. Telomerase-targeting compounds Imetelstat and 6-thio-dG act synergistically with chemotherapy in high-risk neuroblastoma models. Cellular Oncology. 2022;**45**(5):991-1003

[74] Harley CB. Telomerase and cancer therapeutics. Nature Reviews Cancer. 2008;**8**(3):167-179

[75] Kawauchi K, Akiyama M, Yamada O. The mechanisms of telomere and telomerase regulation in hematologic malignancies. Frontiers in Clinical Drug Research - Anti-Cancer Agents. 2013:115-183

[76] Gorecki L et al. Discovery of ATR kinase inhibitor berzosertib (VX-970, M6620): Clinical candidate for cancer therapy. Pharmacology and Therapeutics. 2020;**210**:107518

[77] Laroche-Clary A et al. ATR inhibition broadly sensitizes soft-tissue sarcoma cells to chemotherapy independent of alternative lengthening telomere (ALT) status. Scientific Reports. 2020;**10**(1):7488

[78] Bhatia K, Das A. Combinatorial drug therapy in cancer-new insights. Life Sciences. 2020;**258**:118134

[79] Frederico SC et al. Making a cold tumor hot: The role of vaccines in the treatment of glioblastoma. Frontiers in Oncology. 2021;**11**:672508

[80] Erasimus H et al. DNA repair mechanisms and their clinical impact in glioblastoma. Mutation Research/ Reviews in Mutation Research. 2016;**769**:19-35

[81] Sundar R et al. Targeting ATR in cancer medicine. Current Problems in Cancer. 2017;**41**(4):302-315

[82] De Vitis M, Berardinelli F, Sgura A. Telomere length maintenance in cancer: at the crossroad between telomerase and alternative lengthening of telomeres (ALT). International Journal of Molecular Sciences. 2018;**19**(2):606

[83] Brown JS et al. Targeting DNA repair in cancer: Beyond PARP inhibitors. Cancer Discovery. 2017;7(1):20-37

[84] Brankiewicz W et al. Targeting shelterin proteins for cancer therapy. Drug Discovery Today. 2024:104056

[85] Concannon K et al. Combining targeted DNA repair inhibition and immune-oncology approaches for enhanced tumor control. Molecular Cell. 2023;**83**(5):660-680

[86] Spanakis M, Tsatsakis A. Telomeres, Telomerase, and Shelterin Complex: Promising Drug Targets for Longevity, Age-Related Diseases, and Cancer, in Telomeres. Jenny Stanford Publishing; 2025. pp. 645-667

[87] Wu J, Waxman DJ. Immunogenic chemotherapy: Dose and schedule dependence and combination with immunotherapy. Cancer Letters. 2018;**419**:210-221

[88] Wang X, Zhang H, Chen X. Drug resistance and combating drug resistance in cancer. Cancer Drug Resistance. 2019;**2**(2):141

[89] Liu N et al. Role of telomerase in the tumour microenvironment. Clinical and Experimental Pharmacology and Physiology. 2020;**47**(3):357-364

[90] Carnero A, Lleonart M. The hypoxic microenvironment: A determinant of cancer stem cell evolution. Inside the Cell. 2016;**1**(2):96-105

[91] Robinson NJ, Schiemann WP. Means to the ends: The role of telomeres and telomere processing machinery in metastasis. Biochimica et Biophysica Acta (BBA)-Reviews on Cancer. 2016;**1866**(2):320-329

[92] Liao W et al. Targeting cancer stem cells and signalling pathways through

phytochemicals: A promising approach against colorectal cancer. Phytomedicine. 2023;**108**:154524

[93] Saeed H, Iqtedar M. Stem cell function and maintenance–ends that matter: Role of telomeres and telomerase. Journal of Biosciences. 2013;**38**:641-649

[94] Lupatov AY, Yarygin KN. Telomeres and telomerase in the control of stem cells. Biomedicine. 2022;**10**(10):2335

[95] Haendeler J et al. Mitochondrial telomerase reverse transcriptase binds to and protects mitochondrial DNA and function from damage. Arteriosclerosis, Thrombosis, and Vascular Biology. 2009;**29**(6):929-935

[96] Saretzki G. Extra-telomeric functions of human telomerase: Cancer, mitochondria and oxidative stress. Current Pharmaceutical Design. 2014;**20**(41):6386-6403

[97] Xu Y et al. Emerging roles of the p38 MAPK and PI3K/AKT/mTOR pathways in oncogene-induced senescence. Trends in Biochemical Sciences. 2014;**39**(6):268-276

[98] Augustine T, Maitra R, Goel S. Telomere length regulation through epidermal growth factor receptor signaling in cancer. Genes and Cancer. 2017;**8**(5-6):550

[99] Nieszporek A et al. Molecular mechanisms of epithelial to mesenchymal transition in tumor metastasis. Acta Biochimica Polonica. 2019;**66**(4)

[100] Chakraborty S, Banerjee S. Combatting cellular immortality in cancers by targeting the shelterin protein complex. Biology Direct. 2024;**19**(1):120

[101] Wu Y, Poulos RC, Reddel RR. Role of POT1 in human cancer. Cancers. 2020;**12**(10):2739

[102] Bhari VK et al. Shelterin complex gene: Prognosis and therapeutic vulnerability in cancer. Biochemistry and Biophysics Reports. 2021;**26**:100937

[103] Patel TN et al. Shelterin proteins and cancer. Asian Pacific Journal of Cancer Prevention. 2015;**16**(8):3085-3090

[104] Bilsland AE, Cairney CJ, Nicol Keith W. Targeting the telomere and shelterin complex for cancer therapy: Current views and future perspectives. Journal of Cellular and Molecular Medicine. 2011;**15**(2):179-186

[105] Wang Z, Wu X. Abnormal function of telomere protein TRF2 induces cell mutation and the effects of environmental tumor-promoting factors. Oncology Reports. 2021;**46**(2):1-20

[106] Amir M et al. Structural features of nucleoprotein CST/shelterin complex involved in the telomere maintenance and its association with disease mutations. Cells. 2020;**9**(2):359

[107] Di Vito S. Role of Telomeric Protein TRF2 in Aging and Cancer. 2022

[108] Kallingal A et al. TRF1 and TRF2: Pioneering targets in telomere-based cancer therapy. Journal of Cancer Research and Clinical Oncology. 2024;**150**(7):353

[109] Ng YB, Akincilar SC. Shaping DNA damage responses: Therapeutic potential of targeting telomeric proteins and DNA repair factors in cancer. Current Opinion in Pharmacology. 2024;**76**:102460

[110] Dogan F, Forsyth NR. Telomerase regulation: A role for epigenetics. Cancers. 2021;**13**(6):1213

[111] Jezek M, Green EM. Histone modifications and the maintenance of telomere integrity. Cells. 2019;**8**(2):199

[112] Wu S et al. BRG1, the ATPase subunit of SWI/SNF chromatin remodeling complex, interacts with HDAC2 to modulate telomerase expression in human cancer cells. Cell Cycle. 2014;**13**(18):2869-2878

[113] Bieluszewski T et al. The role and activity of SWI/SNF chromatin remodelers. Annual Review of Plant Biology. 2023;**74**(1):139-163

[114] Guo F et al. Long noncoding RNA: A resident staff of genomic instability regulation in tumorigenesis. Cancer Letters. 2021;**503**:103-109

[115] Aguado J, di Fagagna FdA, Wolvetang E. Telomere transcription in ageing. Ageing Research Reviews. 2020;**62**:101115

[116] Gutschner T, Diederichs S. The hallmarks of cancer: A long non-coding RNA point of view. RNA Biology. 2012;**9**(6):703-719

[117] Nandwani A, Rathore S, Datta M. LncRNAs in cancer: Regulatory and therapeutic implications. Cancer Letters. 2021;**501**:162-171

[118] Sherazi SAM et al. Molecular hallmarks of long non-coding RNAs in aging and its significant effect on aging-associated diseases. Neural Regeneration Research. 2023;**18**(5):959-968

[119] Cacchione S, Biroccio A, Rizzo A. Emerging roles of telomeric chromatin alterations in cancer. Journal of Experimental and Clinical Cancer Research. 2019;**38**(1):21

[120] Li J, Poi MJ, Tsai M-D. Regulatory mechanisms of tumor suppressor P16INK4A and their relevance to cancer. Biochemistry. 2011;**50**(25):5566-5582

[121] Ohtani N et al. The p16INK4a-RB pathway: Molecular link between cellular senescence and tumor suppression. The Journal of Medical Investigation. 2004;**51**(3, 4):146-153

[122] Rowland TJ et al. Single-cell imaging reveals unexpected heterogeneity of telomerase reverse transcriptase expression across human cancer cell lines. Proceedings of the National Academy of Sciences. 2019;**116**(37):18488-18497

[123] Wang F et al. Robust measurement of telomere length in single cells. Proceedings of the National Academy of Sciences. 2013;**110**(21):E1906-E1912

[124] Demanelis K et al. Determinants of telomere length across human tissues. Science. 2020;**369**(6509):eaaz6876

[125] Brane AC, Tollefsbol TO. Targeting telomeres and telomerase: Studies in aging and disease utilizing CRISPR/Cas9 technology. Cells. 2019;**8**(2):186

[126] Jiang Y et al. Detecting telomerase activity at the single-cell level using a CRISPR-Cas12a-based chip. Lab on a Chip. 2025;**25**(1):49-56

[127] Porika M, Tippani R, Saretzki GC. CRISPR/Cas: A new tool in the research of telomeres and telomerase as well as a novel form of cancer therapy. International Journal of Molecular Sciences. 2022;**23**(6):3002

[128] Amorim JPCA. Imaging telomeres by Crispr/Cas9. Portugal: Universidade do Porto; 2016

[129] Holesova Z et al. Telomere length changes in cancer: Insights on carcinogenesis and potential for non-invasive diagnostic strategies. Genes. 2023;**14**(3):715

[130] Pallis AG et al. Evaluating the physiological reserves of older patients

with cancer: The value of potential biomarkers of aging? Journal of Geriatric Oncology. 2014;**5**(2):204-218

[131] Tomasova K et al. Telomere maintenance in interplay with DNA repair in pathogenesis and treatment of colorectal cancer. Mutagenesis. 2020;**35**(3):261-271

[132] Alagoz M et al. DNA repair and resistance to topoisomerase I inhibitors: Mechanisms, biomarkers and therapeutic targets. Current Medicinal Chemistry. 2012;**19**(23):3874-3885

[133] Tao H-Y et al. Targeting telomere dynamics as an effective approach for the development of cancer therapeutics. International Journal of Nanomedicine. 2024:3805-3825

[134] Ali JH, Walter M. Combining old and new concepts in targeting telomerase for cancer therapy: Transient, immediate, complete and combinatory attack (TICCA). Cancer Cell International. 2023;**23**(1):197

[135] Berei J et al. Potential telomere-related pharmacological targets. Current Topics in Medicinal Chemistry. 2020;**20**(6):458-484

[136] Akincilar SC, Unal B, Tergaonkar V. Reactivation of telomerase in cancer. Cellular and Molecular Life Sciences. 2016;**73**:1659-1670

[137] Leão R et al. Mechanisms of human telomerase reverse transcriptase (h TERT) regulation: Clinical impacts in cancer. Journal of Biomedical Science. 2018;**25**:1-12

[138] Hu J et al. Antitelomerase therapy provokes ALT and mitochondrial adaptive mechanisms in cancer. Cell. 2012;**148**(4):651-663

[139] Mills KD, Ferguson DO, Alt FW. The role of DNA breaks in genomic instability and tumorigenesis. Immunological Reviews. 2003;**194**(1):77-95

[140] Lee J-K et al. Mechanisms and consequences of cancer genome instability: Lessons from genome sequencing studies. Annual Review of Pathology: Mechanisms of Disease. 2016;**11**(1):283-312

[141] Svenson U et al. Breast cancer survival is associated with telomere length in peripheral blood cells. Cancer Research. 2008;**68**(10):3618-3623

[142] Rode L, Nordestgaard BG, Bojesen SE. Peripheral blood leukocyte telomere length and mortality among 64 637 individuals from the general population. Journal of the National Cancer Institute. 2015;**107**(6):djv074

[143] Willeit P et al. Telomere length and risk of incident cancer and cancer mortality. JAMA. 2010;**304**(1):69-75

[144] Zhu X et al. The association between telomere length and cancer risk in population studies. Scientific Reports. 2016;**6**(1):22243

[145] Muresanu C et al. Updated understanding of cancer as a metabolic and telomere-driven disease, and proposal for complex personalized treatment, a hypothesis. International Journal of Molecular Sciences. 2020;**21**(18):6521

[146] Ishikawa N et al. Changes of telomere status with aging: An update. Geriatrics and Gerontology International. 2016;**16**:30-42

[147] Gatzka MV. Targeted tumor therapy remixed—An update on the use of small-molecule drugs in combination therapies. Cancers. 2018;**10**(6):155

[148] Zhu L et al. The developing landscape of combinatorial therapies of immune checkpoint blockade with DNA damage repair inhibitors for the treatment of breast and ovarian cancers. Journal of Hematology and Oncology. 2021;**14**:1-19

[149] Thombare K. The Roles of RNA m6A Modification in Disease Mechanisms of High-Risk Neuroblastoma. 2024

[150] Zhang Q et al. Single-cell transcriptome reveals cellular hierarchies and guides p-EMT-targeted trial in skull base chordoma. Cell Discovery. 2022;**8**(1):94

[151] Vodicka P et al. The interactions of DNA repair, telomere homeostasis, and p53 mutational status in solid cancers: Risk, prognosis, and prediction. Cancers. 2021;**13**(3):479

Chapter 5

Ubiquitinomics in Sigmoid Colon Cancer: Defining Stage-Specific Patterns for Predictive, Preventive, and Personalized Medicine

Ousman Bajinka, Lamarana Jallow, Haiyu Shen, Na Li and Xianquan Zhan

Abstract

Sigmoid colon cancer (SCC) is a distinct clinical and molecular subtype of colorectal cancer, characterized by unique molecular, diagnostic, therapeutic, and prognostic features. Recent ubiquitinomics profiling identified 608 differentially ubiquitinated proteins (DUPs) with 1249 ubiquitinated sites, revealing stage-specific patterns. Early stages exhibit ubiquitination changes facilitating oncogenic evasion, while advanced stages show DUPs driving metastasis, therapy resistance, and immune evasion. Key pathways include *Salmonella* infection, ferroptosis, and glycolysis. Crucially, 46 survival-associated DUPs serve as prognostic biomarkers and therapeutic targets. Despite these advances, clinical translation faces limitations, including technical challenges in ubiquitinomics such as inability to resolve polyubiquitin chain topology, biological heterogeneity, subsite generalization, and validation gaps. Future efforts require refined ubiquitin detection methods, functional validation of DUPs, multi-omic integration, and prospective clinical studies. Overcoming these barriers will enable predictive diagnostics, preventive strategies like targeting pre-malignant ubiquitination, and personalized therapies such as DUB inhibitors, PROTACs, advancing the predictive, preventive, personalized medicine (3 PM) framework for SCC management.

Keywords: sigmoid colon cancer, ubiquitinomics, predictive preventive personalized medicine (3 PM), stage-specific biomarkers, therapeutic targeting

1. Introduction

Sigmoid colon cancer (SCC), arising from the distal descending colon proximal to the rectum, represents a significant and distinct clinical entity within colorectal cancer (CRC). Epidemiologically, it is the second most common CRC site, accounting for 16–31% of cases and is generally associated with a better prognosis than right-sided colon cancer, though it paradoxically correlates with a higher incidence of suicide, underscoring its unique clinical profile [1–4]. Molecularly, SCC is characterized by specific alterations that differentiate it from other colorectal subsites. A defining feature is its unique ubiquitinomic signature, involving 608 differentially ubiquitinated proteins linked to pathways such as *Salmonella* infection and ferroptosis [5]. This profile contrasts with right-sided tumors, which often exhibit suppression of carcinogen-detoxifying enzymes [6]. Further molecular distinctions include epigenetic alterations like frequent *p16INK4a* methylation in aggressive subsets, specific mutations such as *KRAS* G12C in rare amphicrine carcinomas, and markers of epithelial-mesenchymal transition (EMT) like E-cadherin loss [7–9].

Diagnostic approaches for SCC have advanced significantly, with MRI demonstrating superiority over CT for local T-staging and detection of extramural vascular invasion [10]. These imaging techniques are complemented by emerging liquid biopsies, including 7-gene panels and circulating tumor DNA (ctDNA) for predicting micrometastasis and recurrence, as well as the analysis of exfoliated colonocytes [11–14]. Therapeutically, management strategies reflect the distinct anatomy and behavior of SCC. Robotic surgery enables precise D3-D4 lymphadenectomy for advanced nodal disease, while preservation of the superior rectal artery (SRA) can

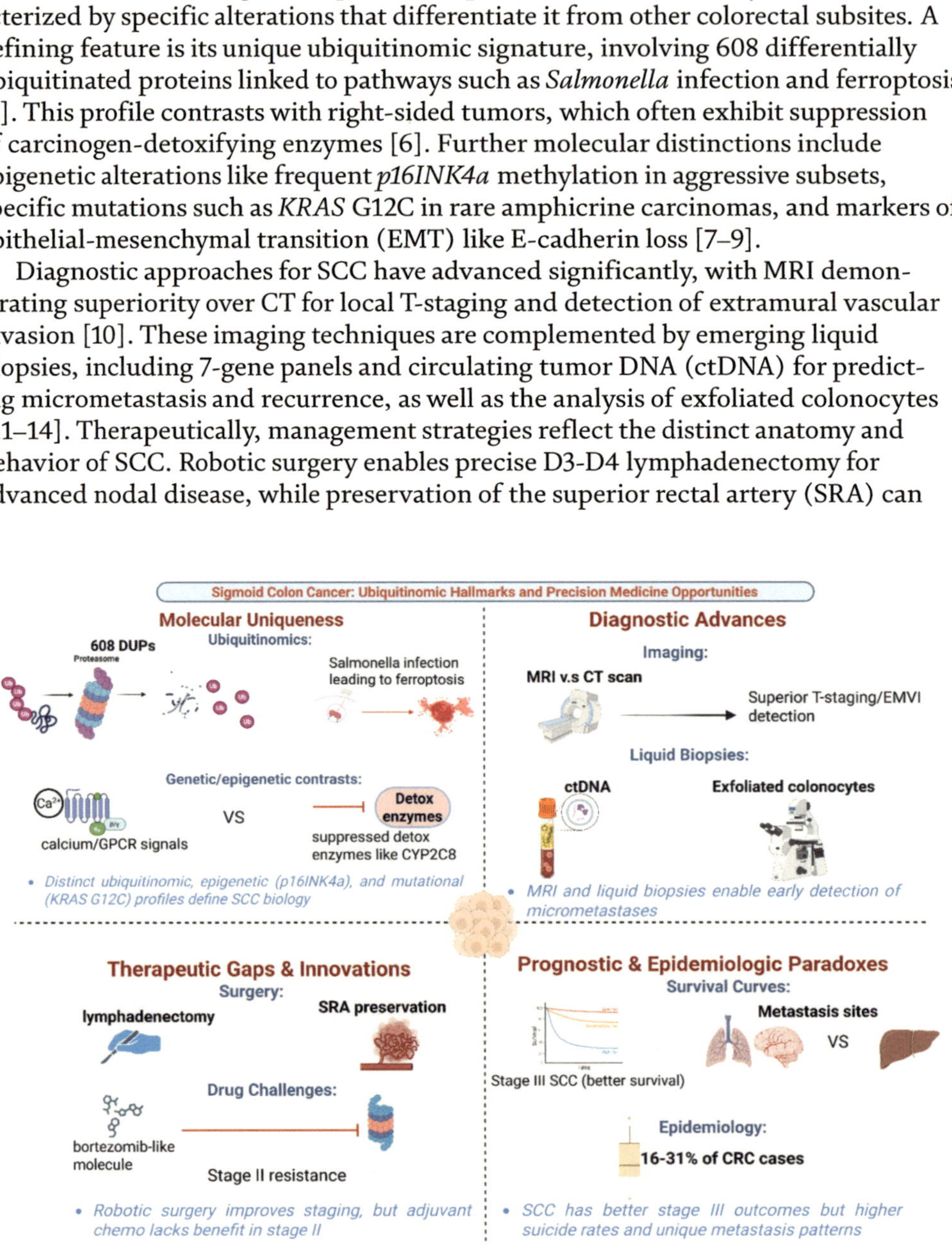

Figure 1.
Key ubiquitinomic features and precision medicine opportunities in sigmoid colon cancer (SCC). This schematic highlights the major molecular, diagnostic, therapeutic, and prognostic features of sigmoid colon cancer (SCC) with relevance to precision medicine. Top-left: SCC displays unique ubiquitinomic signatures, including 608 differentially ubiquitinated proteins (DUPs), links to Salmonella-induced ferroptosis, and distinct genetic/epigenetic alterations such as KRAS G12C mutations, p16/INK4A loss, altered Ca^{2+}/GPCR signaling, and reduced detoxification enzymes (CYP2C8). Advancements in diagnosis include improved MRI vs. CT accuracy for T-staging/EMVI and the utility of liquid biopsies (ctDNA and exfoliated colonocytes) for early micrometastatic detection. Bottom-left: Therapeutic gaps include limited benefits of adjuvant chemotherapy in stage II SCC, drug resistance to bortezomib-like inhibitors, and surgical considerations such as lymphadenectomy vs. SRA preservation. Bottom-right: Prognostic patterns show better stage III survival, distinct metastatic spread (brain/lung > liver), and notable epidemiologic burdens, with SCC representing 16–31% of CRC cases.

reduce anastomotic complications [15, 16]. For localized tumors invading adjacent organs, en bloc resection is critical for local control. However, adjuvant chemotherapy shows no clear benefit in stage II disease, and the use of self-expanding metal stents (SEMS) for palliation carries high complication risks [17–20].

Prognostically, the sigmoid location itself is an independent factor, conferring better outcomes in stage III cancer and after liver metastasis development compared to proximal sites [21–23]. Nevertheless, SCC is associated with higher rates of lung and brain metastases, and its prognosis is further refined by lymph node distribution and molecular subtypes, including MSI status and markers like CPEB4 expression [24–29]. Despite these insights, current research faces significant limitations, including subsite generalization, small cohorts for rare subtypes, retrospective data biases, and a lack of large-scale validation for novel diagnostic and prognostic tools. Therefore, this chapter aims to synthesize the distinct molecular, diagnostic, therapeutic, and prognostic characteristics of SCC, with a primary focus on establishing disease- and stage-specific ubiquitination patterns as novel hallmarks. It critically evaluates existing research limitations while highlighting the implications of these ubiquitination signatures for advancing predictive, preventive, and personalized medicine (3 PM) in SCC management.

The ubiquitinomics profiling was conducted using a cohort of 100 SCC samples and matched para-carcinoma tissues from The Cancer Genome Atlas (TCGA). The methodology involved the enrichment of ubiquitinated peptides using anti-K-ε-GG antibody beads, followed by label-free quantitative high-resolution mass spectrometry to identify and quantify differentially ubiquitinated proteins (DUPs) and their specific modification sites. This chapter aims to synthesize current knowledge on the distinct molecular, diagnostic, therapeutic, and prognostic characteristics of SCC, with a primary focus on establishing *disease-* and *stage-specific ubiquitination patterns as novel* hallmarks and evaluating their implications for advancing 3 PM in SCC management (**Figure 1**).

2. Establishment of disease and stage-specific ubiquitination patterns for human SCC

Ubiquitination, the covalent attachment of ubiquitin molecules to target proteins, is a fundamental post-translational modification (PTM) regulating virtually every cellular process, including protein degradation (via the proteasome or lysosome), DNA repair, signal transduction, immune responses, and cell cycle progression. Dysregulation of the ubiquitin-proteasome system (UPS) is a hallmark of cancer, contributing to oncogenesis, tumor progression, metastasis, and treatment resistance. While global proteomic and genomic landscapes of CRC have been extensively studied, the specific role of ubiquitination, particularly in a subsite- and stage-specific manner, remained largely unexplored until recently. The establishment of disease and stage-specific ubiquitination patterns in human SCC represents a significant advancement in our molecular understanding of this prevalent malignancy, offering novel insights for predictive, preventive, and personalized medicine (3 PM) [5].

2.1 The ubiquitinomics approach: Unveiling the sigmoideome

The pioneering study by Yang et al. [5] established the first comprehensive ubiquitinomic profile specific to human SCC by employing a sophisticated multi-faceted methodology: utilizing a cohort of 100 SCC samples and matched para-carcinoma tissues from The Cancer Genome Atlas (TCGA) to ensure robust clinical annotation;

leveraging anti-K-ε-GG antibody beads (PTMScan® Ubiquitin Remnant Motif Kit) to enrich tryptic peptides containing the diglycine (Gly-Gly) remnant left on lysine residues after ubiquitination, thereby isolating peptides representing ubiquitinated sites; performing label-free quantitative high-resolution mass spectrometry (MS) on these enriched peptides to identify and quantify ubiquitinated proteins and their specific modification sites; and integrating this ubiquitinome data with corresponding gene expression (differentially expressed genes – DEGs) and proteomic (differentially expressed proteins – DEPs) data from the same cohort to explore relationships between ubiquitination, gene/protein expression, and protein abundance, supplemented by Kaplan-Meier survival analysis to identify prognostic associations.

2.2 Establishing the baseline: The sigmoid colon ubiquitinome

This groundbreaking work quantified a vast and complex ubiquitination landscape specific to SCC tissues compared to adjacent normal tissue, revealing *1249 ubiquitinated sites* within *608 differentially ubiquitinated proteins (DUPs)*. This scale highlights ubiquitination's pervasive role as a fundamental regulatory mechanism in SCC biology. A critical advancement was the stratification of these DUPs by disease progression. Analysis of the TCGA cohort revealed dynamic, stage-specific patterns: *early-stage SCC was characterized by 385 DUPs* often involved in processes like metabolic reprogramming and evasion of growth suppression, while *advanced-stage SCC exhibited a distinct set of 223 DUPs* driving metastasis, therapy resistance, and immune evasion.

Here, DUPs are defined as proteins exhibiting statistically significant differences in ubiquitination levels (quantified by ubiquitinated peptide abundance) between cancerous and normal tissues. This encompasses both *increased ubiquitination (DUP-up)*, often leading to target protein degradation or functional alteration, and *decreased ubiquitination (DUP-down)*, potentially stabilizing oncoproteins or dysregulating critical signaling pathways. Most significantly, the prognostic relevance of this ubiquitinome was rigorously tested. Survival analysis, specifically Kaplan-Meier methodology applied to the patient cohort, identified *46 overall survival-related DUPs*. The ubiquitination status of these specific proteins showed a statistically significant association with patient outcomes, thereby validating their potential as prognostic biomarkers and implicating them as hallmarks of aggressive, high-stage biology.

2.3 Stage-specific ubiquitination patterns emerge

A critical finding was the identification of ubiquitination patterns correlating with SCC progression stage, observed through analysis of the TCGA cohort spanning different stages. The study revealed dynamic shifts in ubiquitination status across protein networks: early stages showed altered ubiquitination (often increased degradation) in proteins maintaining cellular integrity, DNA repair, and controlled metabolism, facilitating initial oncogenic evasion, while advanced stages exhibited distinct DUP patterns associated with metastatic potential, involving changes in cell adhesion (integrins and cadherins), cytoskeletal dynamics, and EMT markers [9], potentially via upregulated deubiquitinating enzymes (DUBs) stabilizing EMT transcription factors; treatment resistance, through alterations in ubiquitination pathways controlling drug efflux pumps, DNA repair machinery, and apoptosis regulators (Bcl-2 family, IAPs); and immune evasion, via modulation of antigen presentation (MHC), immune checkpoint stability (PD-L1 regulated by E3 ligases like CBL-b or DUBs), and cytokine signaling.

KEGG pathway analysis of the 608 DUPs implicated 35 significant pathways, including *Salmonella* infection (reflecting hijacking of host ubiquitination for invasion/metastasis), glycolysis/gluconeogenesis (stage-specific regulation of key enzymes like PKM2/LDHA controlling Warburg effect metabolic reprogramming), and ferroptosis (ubiquitination dynamics of regulators like GPX4/SLC7A11 influencing susceptibility to this iron-dependent cell death, potentially altered in advanced disease via DUB upregulation). Most significantly, survival analysis identified 46 overall survival-related DUPs whose ubiquitination status directly correlated with patient outcomes, strongly implicating these events as hallmarks of aggressive, high-stage biology, and prime candidates for stage-specific biomarkers.

2.4 Functional implications of stage-specific ubiquitination

The observed ubiquitination patterns are not merely correlative but drive SCC pathogenesis through stage-dependent functional consequences: Oncoprotein stabilization (DUP-down) occurs via decreased ubiquitination preventing the degradation of key oncoproteins like β-catenin (enhancing Wnt signaling) [9], c-Myc, cyclin D1 (accelerating proliferation), or RTKs such as EGFR/MET, hyperactivating growth pathways; tumor suppressor degradation (DUP-up) involves increased ubiquitination-mediated proteasomal destruction of p53 (abrogating apoptosis), PTEN (activating PI3K/AKT survival signals), and adhesion/DNA repair regulators (fueling metastasis/genomic instability); altered signaling dynamics arise from direct ubiquitination of components in critical pathways, MAPK/ERK (RAS/RAF), PI3K/AKT/mTOR (AKT/mTOR), NF-κB (IκBs/NEMO controlling therapy resistance), and TGF-β (receptors/SMADs modulating EMT in advanced stages) [9]. Moreover, metabolic reprogramming results from ubiquitination controlling enzyme stability (locking in Warburg effect via PKM2/LDHA regulation). Immune evasion exploits ubiquitination-mediated modulation of immune checkpoints (PD-L1 stability), antigen presentation, and cytokine responses to create an immunosuppressive microenvironment in advanced SCC.

2.5 Relationship to gene and protein expression (DEGs/DEPs)

A unique aspect of the Yang et al. study [5] was the integrated analysis of the ubiquitinome with DEGs and protein expression (DEPs), revealing four distinct regulatory relationship models: (1) DUP-up and DEG-up (increased ubiquitination and gene expression, suggesting futile cycling or non-degradative ubiquitin signaling); (2) DUP-up and DEG-down (increased ubiquitination with decreased gene expression, indicating canonical UPS-mediated degradation); (3) DUP-down and DEG-up (decreased ubiquitination and increased gene expression, implying protein stabilization driving oncogenic accumulation); and (4) DUP-down and DEG-down (decreased ubiquitination despite reduced gene expression, potentiating functional impact of stabilized residual proteins). Similar quadripartite models emerged for DUP-DEP relationships, collectively demonstrating that ubiquitination constitutes an independent regulatory layer beyond transcriptional or translational control. Critically, stage-specific cancer progression likely involves dynamic shifts in the prevalence of these models, for example, advanced SCC may exhibit increased "DUP-down and DEG-up" stabilization of oncogenes (*β-catenin, c-Myc*) and "DUP-up and DEG-down" degradation of tumor suppressors (*p53, PTEN*), amplifying oncogenic signaling through coordinated multi-level dysregulation.

2.6 Implications for 3 PM

The establishment of disease and stage-specific ubiquitination patterns is foundational for advancing 3 PM in SCC, offering transformative potential across the clinical spectrum: predictive diagnostics leverage the 46 survival-related DUPs and other stage-specific signatures as prognostic biomarkers, where detecting specific ubiquitination patterns (in biopsies or future liquid biopsies) could stratify patients by progression/recurrence risk at diagnosis, guiding treatment intensity; preventive measures target the earliest ubiquitination changes in high-risk individuals (IBD patients, genetic syndromes), enabling the development of chemopreventive agents modulating specific E3 ligases or DUBs active in pre-malignant stages; and personalized therapy exploits these patterns for novel interventions, including targeting specific E3 ligases (MDM2 inhibitors for p53-restorable tumors), inhibiting oncoprotein-stabilizing DUBs (USP7 inhibitors in advanced SCC), refining proteasome inhibitors (targeting SCC-specific vulnerabilities like immunoproteasomes), designing rational combination therapies (DUB inhibitors + targeted agents based on

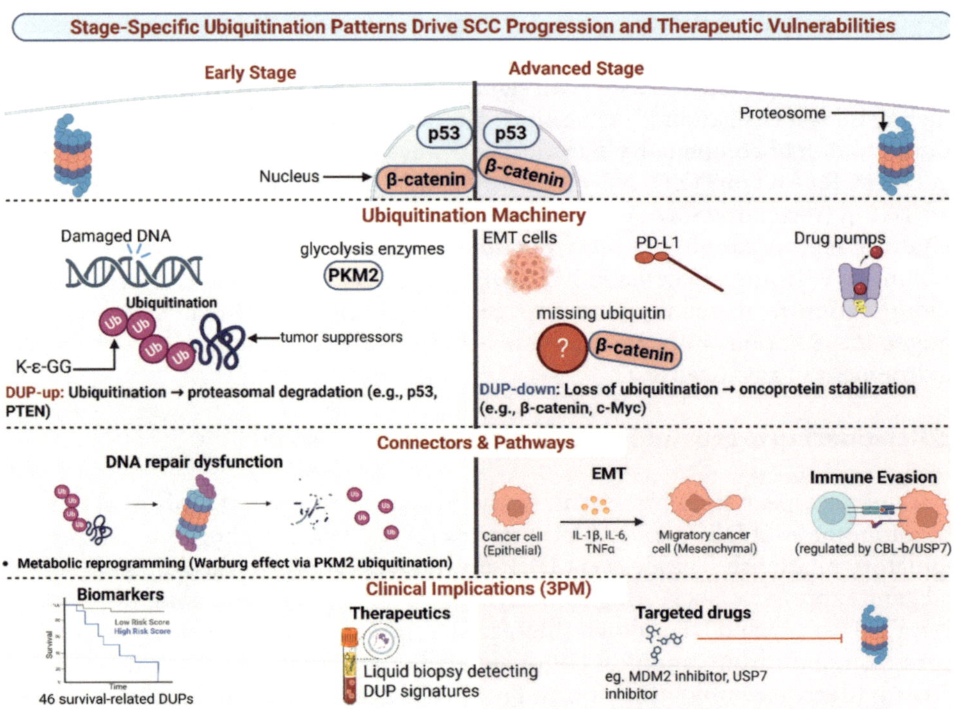

Figure 2.
Stage-specific ubiquitination patterns drive SCC progression and therapeutic vulnerabilities. This figure illustrates how dynamic ubiquitination patterns shape sigmoid colon cancer (SCC) biology across disease stages and reveal precision-medicine targets. In early-stage SCC, increased ubiquitination (DUP-up) promotes proteasomal degradation of tumor suppressors (e.g., p53, PTEN) and metabolic reprogramming through PKM2 ubiquitination, contributing to DNA repair dysfunction and enhanced glycolysis. In advanced-stage SCC, reduced ubiquitination (DUP-down) leads to stabilization of oncogenic proteins (e.g., β-catenin, c-Myc), supporting epithelial-mesenchymal transition (EMT), drug resistance via efflux pumps, and immune evasion (e.g., PD-L1 signaling regulated by CBL-b/USP7). Key connectors include inflammatory cytokines driving EMT (IL-1β, IL-6, TNFα). Clinical implications (3 PM) include (i) 46 survival-associated DUPs as prognostic biomarkers, (ii) liquid biopsy-based detection of DUP signatures, and (iii) stage-tailored therapeutics such as MDM2 or USP7 inhibitors targeting ubiquitin-regulating enzymes.

ubiquitination-mediated resistance), and leveraging drug sensitivity profiling linked to survival-related DUPs to identify personalized therapeutic vulnerabilities [5]. This integrated approach enables precise risk stratification, interception of carcinogenesis, and tailored targeting of the ubiquitin-driven molecular machinery specific to SCC biology.

The establishment of disease and stage-specific ubiquitination patterns for human SCC, pioneered by the ubiquitinomics approach of Yang et al. [5], marks a paradigm shift in understanding this common malignancy. The identification of 608 DUPs with 1249 ubiquitinated sites, their association with critical pathways (*Salmonella* infection, glycolysis, and ferroptosis), their complex relationships with gene and protein expression, and crucially, the discovery of 46 survival-related DUPs, provides an unprecedented molecular map of SCC progression. These patterns reveal how the UPS is hijacked at different stages to stabilize oncoproteins, degrade tumor suppressors, rewire signaling, reprogram metabolism, and evade the immune system. While significant challenges in validation, functional characterization, and clinical translation remain, this foundational work opens powerful avenues for advancing predictive diagnostics, preventive strategies, and personalized therapies tailored to the specific ubiquitination-driven vulnerabilities of an individual's SCC, ultimately realizing the promise of 3 PM. Future research must focus on overcoming technical limitations, resolving spatial heterogeneity, performing deep functional validation, integrating multi-omic data, and rigorously translating these discoveries into clinical tools and treatments that improve patient outcomes (**Figure 2**; **Table 1**).

Study objective	Study design	Main findings	Conclusion	Citation
Evaluate robotic D3-D4 lymphadenectomy for advanced nodal disease	Single-center clinical study	Improved nodal staging accuracy and survival in advanced sigmoid/rectal cancer	Robotic lymphadenectomy enhances precision and survival in advanced disease	[15]
Compare MRI vs. CT for local staging	Diagnostic accuracy study	MRI superior to CT in T-staging and detecting extramural vascular invasion	MRI should replace CT for local staging of SCC	[10]
Assess SRA preservation outcomes	Observational cohort study	Reduced anastomotic leak rates (7 vs. 22% in controls)	SRA preservation lowers surgical complications in sigmoid resection	[16]
Evaluate SEMS for obstruction palliation	Retrospective multicenter study	High complication rates (perforation: 15%, migration: 12%)	SEMS carries significant risks; alternative palliation methods preferred	[20]
Validate ctDNA for recurrence prediction	Prospective case-control study	ctDNA detected micrometastases 6 months earlier than imaging	ctDNA enables early recurrence intervention in stage III–IV SCC	[12]
Compare mini-probe US vs. CT staging	Diagnostic validation study	Mini-probe US accuracy: 92% vs. CT: 74% for T-stage	Endoscopic ultrasound outperforms CT in local staging accuracy	[13]

Table 1.
Summary of key studies on SCC management.

3. Limitations in establishing disease and stage-specific ubiquitination patterns for human SCC

The groundbreaking work by Yang et al. [5] establishing the first ubiquitinomic profile specific to human SCC represents a significant leap forward. However, translating these findings into clinically actionable tools for predictive, 3 PM faces substantial limitations across technical, biological, and clinical domains:

3.1 Technical complexity and limitations of ubiquitinomics

Ubiquitinomics faces substantial methodological challenges due to its technical complexity and resource intensity. The core technique, enriching K-ε-GG peptides via antibodies followed by high-resolution mass spectrometry (MS), is highly sensitive to sample quality, digestion efficiency, and enrichment biases, with poor cross-laboratory reproducibility [5]. Crucially, the anti-K-ε-GG antibody captures ubiquitin remnants but fails to differentiate between mono-ubiquitination and functionally distinct poly-ubiquitin chain linkages (K48-linked for degradation, K63-linked for signaling, M1-linked for NF-κB activation), leaving chain topology, a critical determinant of functional consequences, largely uncharacterized in SCC [5]. Accurate quantification is hindered by limitations of label-free methods (LFQ) for low-abundance peptides, where dynamic range constraints obscure changes in biologically significant DUPs; isobaric tagging (TMT) mitigates this but introduces ratio compression artifacts. Furthermore, ubiquitination operates within a network of PTMs, interacting extensively with phosphorylation, acetylation, and SUMOylation, yet current single-PTM focused methodologies cannot capture this regulatory crosstalk essential to SCC biology [5].

3.2 Subsite generalization and cohort limitations

Despite focusing on SCC, analytical constraints often lead to grouping with other left-sided sites (descending colon) in pathway and survival analyses, potentially obscuring true SCC-specific ubiquitination signatures due to cohort size limitations or study design [5, 6]. While the ubiquitinome cohort (n = 100 tumors) is substantial, it remains underpowered to detect patterns in rare aggressive subtypes like amphicrine carcinoma [8] or molecular subgroups (KRAS/BRAF mutants). Reliance on retrospective TCGA data introduces selection bias and treatment heterogeneity [5], compounded by scarce high-quality fresh-frozen samples with comprehensive clinical annotation for validation. Furthermore, demographic homogeneity in TCGA and similar cohorts [15] limits insights into ethnicity-, geography-, or lifestyle-dependent variations in ubiquitination, restricting global generalizability of findings.

3.3 Biological heterogeneity and functional validation gap

Bulk tissue analysis homogenizes ubiquitination signals across malignant cells, stroma, and normal epithelium, obscuring cell-type-specific dynamics, such as cancer cell-intrinsic immune evasion mechanisms diluted by lymphocyte signatures [5]. This approach further fails to resolve spatial heterogeneity, where ubiquitination patterns likely vary between tumor subregions (invasive front vs. core) and clonal populations. Critically, the identified correlations lack causal validation: the functional roles of specific DUPs (the 46 survival-associated sites) in SCC pathogenesis remain unproven, requiring extensive *in vitro* (E3/DUB modulation) and *in vivo* (PDX models) interrogation [5]. Moreover, ubiquitination's rapid dynamics in response to stimuli (hypoxia and therapy) are lost in static resection snapshots, masking adaptive responses crucial to stage progression and treatment resistance.

3.4 Diagnostic and therapeutic translation hurdles

Translating the 46 survival-related DUPs into clinical biomarkers faces formidable hurdles: current IHC cannot reliably detect ubiquitination patterns in FFPE biopsies, while developing targeted MS or proximity assays remains challenging, and liquid biopsy applications are even more complex [5]. Although advanced-stage ubiquitination signatures (metastasis/resistance/immune evasion) are identified [5], pinpointing druggable stage-specific drivers (E3 ligases/DUBs) requires deeper functional validation. Therapeutic targeting is complicated by specificity issues, off-target effects may disrupt normal UPS function (neurotoxicity), and rapid tumor resistance through compensatory mechanisms. Rational combination strategies (DUB inhibitors + targeted agents) demand better understanding of SCC-specific ubiquitination-pathway crosstalk [5], yet relevant preclinical models recapitulating dynamic SCC ubiquitinomes during progression are currently lacking, impeding drug testing.

3.5 Integration and data interpretation challenges

Comprehensive multi-omic integration, beyond DEGs/DEPs, requires incorporating genomics (*FBXW7*/*TP53* mutations), transcriptomics (ubiquitination-altering splice variants), phosphoproteomics (PTM crosstalk), and metabolomics to build causal stage-specific networks, posing immense computational and biological complexity [5]. Crucially, distinguishing pathogenic *driver* ubiquitination events (direct oncogenic activators) from inconsequential *passenger* changes demands sophisticated bioinformatic prioritization (network centrality metrics) coupled with functional validation, which remains incomplete. Furthermore, clinically actionable thresholds for significant ubiquitination level changes in specific DUPs are undefined, and quantifying the biological relevance of, for instance, a 1.5-fold ubiquitination increase in early-stage SCC is essential for developing diagnostic/prognostic assays but currently lacking.

3.6 Prognostic application limitations

Observed survival advantages linked to distal location or ubiquitination signatures may reflect lead-time bias (earlier diagnosis from symptomatic presentation) or stage distribution confounders rather than intrinsic biological effects of ubiquitination itself [22, 23]. While the 46 DUPs show prognostic potential, their independent value must be validated against established clinical/molecular factors, including TNM stage, CEA levels, MSI status, CMS subtypes, and RAS/BRAF mutations, to determine true incremental predictive power beyond current standards. Most critically, these signatures require rigorous assessment in large, multi-center prospective SCC cohorts prior to clinical implementation, as retrospective data cannot confirm real-world utility for recurrence risk stratification or treatment guidance.

The establishment of disease and stage-specific ubiquitination patterns in SCC is a foundational achievement. However, significant limitations impede its immediate translation into 3 PM. Technical hurdles in ubiquitinomics, biological complexities like heterogeneity and lack of functional validation, challenges in biomarker and

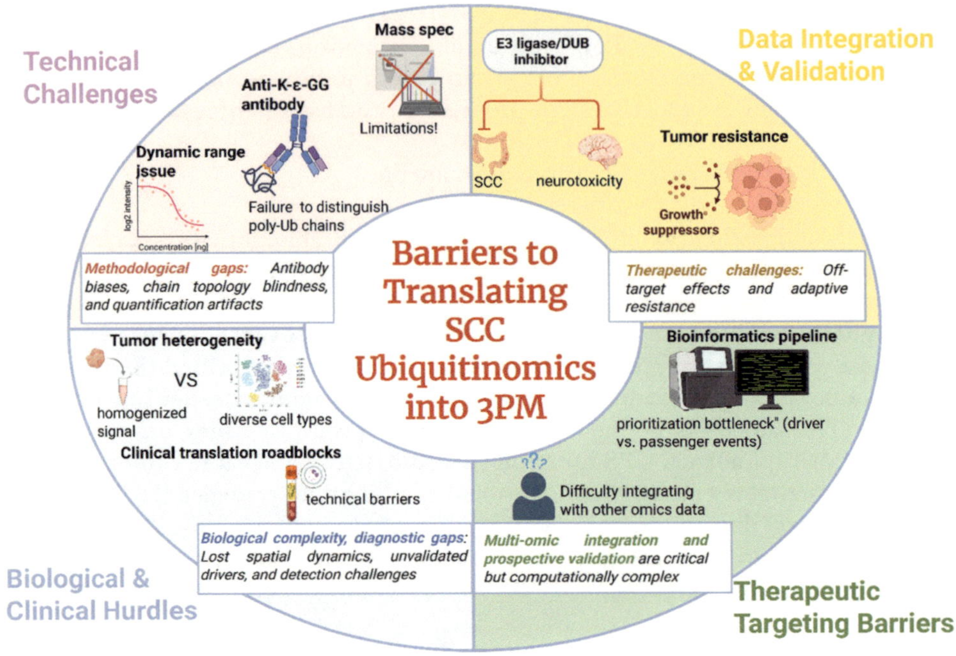

Figure 3.
Barriers to translating SCC ubiquitinomics into predictive, preventive and personalized medicine (3 PM). This figure outlines the major obstacles hindering the clinical translation of Sigmoid Colon Cancer (SCC) ubiquitinomics into 3 PM-driven applications. Technical challenges include mass spectrometry limitations, dynamic range issues, and antibody-based biases such as chain-topology blindness and difficulty distinguishing poly-ubiquitin linkages, leading to quantification artifacts. Biological and clinical hurdles arise from tumor heterogeneity, loss of spatial information, unvalidated ubiquitin-driven events, and limited diagnostic tools, compounded by barriers in moving ubiquitinomics into routine clinical workflows. Data integration and validation barriers include prioritization bottlenecks within bioinformatics pipelines (driver vs. passenger ubiquitination events) and limited prospective, multi-omic validation frameworks. Therapeutic targeting barriers encompass off-target toxicity of E3 ligase/DUB inhibitors, adaptive tumor resistance, and neurotoxicity risks, making targeted drug development and translation challenging.

Pathway	Role/context in SCC	Key mechanisms/molecular components
Salmonella infection	Hijacked for invasion and metastasis; reflects pathogen mimicry of cancer cell invasion.	Utilizes host ubiquitination machinery for intracellular survival and cytoskeletal remodeling, facilitating tumor cell invasion.
Ferroptosis	Iron-dependent cell death pathway; susceptibility is regulated by ubiquitination dynamics, often altered in advanced disease.	Ubiquitination of key regulators like GPX4 and SLC7A11 (system Xc-), influencing redox balance and cell death susceptibility.
Glycolysis/ gluconeogenesis	Drives the Warburg effect (aerobic glycolysis) for metabolic reprogramming; stage-specifically regulated.	Ubiquitination controls the stability and activity of key enzymes such as PKM2 and LDHA, locking in the glycolytic phenotype.
Epithelial-mesenchymal transition (EMT)	Critical for metastasis in advanced stages; driven by ubiquitination-mediated protein stabilization.	Decreased ubiquitination (DUP-down) stabilizes oncoproteins like β-catenin and EMT transcription factors, promoting invasion.

Pathway	Role/context in SCC	Key mechanisms/molecular components
Immune evasion	Creates an immunosuppressive microenvironment in advanced SCC.	Ubiquitination modulates the stability of immune checkpoints like PD-L1 (via E3 ligases e.g., CBL-b or DUBs) and antigen presentation (MHC).
Apoptosis and therapy resistance	Enables evasion of cell death and confers resistance to treatments.	Altered ubiquitination of apoptosis regulators (e.g., Bcl-2 family, IAPs) and DNA repair machinery proteins.
Wnt/β-catenin signaling	Promotes proliferation and stemness; a hallmark oncogenic pathway in SCC.	DUP-down of β-catenin prevents its degradation, leading to constitutive activation of Wnt target genes.
PI3K/AKT/mTOR signaling	Central cell survival pathway; hyperactivated in SCC.	DUP-up of PTEN leads to its degradation, activating PI3K/AKT signals. Direct ubiquitination of AKT/mTOR components also alters signaling dynamics.

This table consolidates the major pathways discussed, highlighting how the ubiquitin-proteasome system (UPS) exerts stage-specific control over critical cancer hallmarks in SCC, from metabolic reprogramming and metastasis to immune evasion and treatment resistance.

Table 2.
Key pathways implicated by ubiquitinomics in sigmoid colon cancer.

Protein	Function	Ubiquitination change in SCC	Implication for 3 PM
β-catenin	Oncoprotein; key mediator of Wnt signaling	DUP-down (Stabilized)	Predictive: Marker of aggressive disease. *Personalized therapy*: Target for Wnt pathway inhibitors or DUB inhibitors.
p53	Tumor suppressor; apoptosis and DNA repair	DUP-up (degraded)	*Predictive:* Poor prognosis biomarker. *Personalized therapy*: Candidate for MDM2 E3 ligase inhibitors in p53-wildtype tumors.
PTEN	Tumor suppressor; negative regulator of PI3K/AKT pathway	DUP-up (degraded)	*Predictive*: Indicator of activated PI3K/AKT signaling and potential therapy resistance. *Personalized therapy*: Guide for AKT/mTOR pathway inhibitors.
PD-L1	Immune checkpoint ligand	DUP-down (stabilized)	*Predictive*: Identifies tumors with enhanced immune evasion potential. *Personalized therapy*: Combine immunotherapy with specific DUB inhibitors that stabilize PD-L1.
c-Myc	Oncogenic transcription factor; drives proliferation	DUP-down (stabilized)	*Predictive*: Marker of high proliferative index. *Personalized therapy*: Target for BET inhibitors or specific E3 ligase agonists.
GPX4	Key regulator of ferroptosis	DUP-up (degraded)	*Predictive*: May indicate increased susceptibility to ferroptosis inducers. *Personalized therapy*: Stratify patients for treatment with ferroptosis-inducing agents.

Note: This table provides representative examples. The full set of 46 DUPs encompasses a wider range of proteins involved in various hallmarks of cancer. Their collective signature offers a powerful multi-dimensional biomarker for 3 PM.

Table 3.
Selected examples of the 46 survival-associated DUPs in sigmoid colon cancer.

therapeutic development, and the need for sophisticated data integration and prospective validation represent major roadblocks. Overcoming these limitations requires concerted efforts: developing more sensitive and specific ubiquitin detection methods (especially for chain topology), employing spatial proteomics and single-cell ubiquitinomics, performing deep functional validation of candidate DUPs in relevant SCC models, integrating ubiquitinomics fully with other omics layers, conducting robust prospective biomarker validation studies, and designing highly specific UPS-targeting agents. Only by addressing these limitations can the immense potential of the SCC ubiquitinome be fully realized for improving patient outcomes through precise prediction, prevention, and personalized therapy (1) (**Figure 3**, **Tables 2** and **3**).

4. Future prospects for SCC management

The establishment of disease- and stage-specific ubiquitination patterns in SCC marks a paradigm shift, moving beyond static genomic and proteomic landscapes to a dynamic, regulatory layer governing tumor behavior. While significant challenges remain, this foundational work illuminates a compelling path forward for predictive, 3 PM in SCC. The future prospects hinge on overcoming current limitations and strategically leveraging the ubiquitinome's potential across several interconnected domains.

4.1 Technical evolution: Decoding the ubiquitin code with precision

The immediate future demands revolutionary advancements in ubiquitinomics technology to overcome the critical bottleneck of accurately detecting polyubiquitin chain topology (distinguishing K48, K63, and M1 linkages), necessitating the development of new generations of topology-specific antibodies or engineered affinity reagents designed to specifically enrich peptides carrying distinct chain linkages, thereby functionally annotating DUPs by distinguishing degradation signals from activation signals or immune modulation. Concurrently, rapid evolution in advanced mass spectrometry (MS) platforms is expected through the integration of trapped ion mobility spectrometry (TIMS), parallel accumulation-serial fragmentation (PASEF), and targeted MS/MS (PRM and SRM) to significantly enhance sensitivity, specificity, and quantification accuracy for low-abundance ubiquitinated peptides, especially from challenging clinical samples like FFPE tissues or liquid biopsies. Furthermore, resolving tumor heterogeneity requires pioneering spatial and single-cell ubiquitinomics techniques, such as adapting imaging mass cytometry for ubiquitin remnants or coupling single-cell proteomics/scRNA-seq with ubiquitin capture, which will reveal cancer cell-intrinsic ubiquitination dynamics versus stromal contributions and map spatial variations within tumor subregions (invasive front versus core); finally, comprehensive understanding of the PTM crosstalk driving SCC progression and therapeutic resistance hinges on developing robust methods for multi-PTM integration, enabling simultaneous or sequential analysis of ubiquitination alongside phosphorylation, acetylation, and SUMOylation via techniques like sequential immunoaffinity purification or multiplexed MS workflows.

4.2 Functional validation: From correlation to causation

The 608 identified DUPs and 46 survival-associated sites constitute a rich reservoir of candidates demanding rigorous functional validation to elucidate their mechanistic roles in SCC pathogenesis, necessitating *high-throughput CRISPR-Cas9 screens* targeting E3 ligases, DUBs, or specific DUP genes in SCC cell lines and organoids to pinpoint essential drivers of oncogenesis, metastasis, and therapy resistance linked to ubiquitination changes; *advanced preclinical models* including patient-derived xenografts (PDXs) and genetically engineered organoids (expressing ubiquitination-resistant oncoprotein mutants or overexpressing DUBs) for *in vivo* validation of stage-specific DUP functions and testing of targeted interventions, with urgent need for models capturing dynamic ubiquitination shifts during therapy; and detailed *mechanistic dissection* through biochemical and cellular studies to decode how specific ubiquitination events (on β-catenin, p53, and PD-L1) alter protein stability, localization, interactions, and downstream signaling cascades (Wnt, PI3K/AKT, and immune checkpoints) within the unique biological context of SCC.

4.3 Biomarker translation: Ubiquitination signatures in the clinic

The immense clinical promise of the 46 survival-related DUPs and stage-specific ubiquitination signatures hinges on overcoming translational barriers, starting with *developing robust assays for routine clinical samples*, including targeted MS panels for key DUP peptides in FFPE biopsies, ubiquitination-specific proximity ligation assays (PLA) for *in situ* chain topology detection, and refined methods for capturing ubiquitin remnants from liquid biopsies (circulating ubiquitinated proteins or tumor exosomes). This must be coupled with *large-scale prospective validation* in dedicated SCC cohorts to rigorously establish the prognostic/predictive power of these signatures against existing markers (TNM, CEA, MSI, and RAS/BRAF) and define clinically actionable thresholds for ubiquitination changes. Furthermore, exploring these signatures for *early detection and MRD monitoring*, such as identifying pre-malignant ubiquitination shifts in blood/stool of high-risk patients (IBD) or enhancing ctDNA analysis with ubiquitination-associated mutational/expression profiles, could revolutionize screening paradigms and residual disease surveillance.

4.4 Therapeutic exploitation: Targeting the ubiquitin system in SCC

The stage-specific ubiquitination map of SCC provides a blueprint for developing novel, personalized therapies. Key strategies include: Precision targeting of the ubiquitin machinery, such as developing or repurposing E3 ligase inhibitors or activators (e.g., MDM2 inhibitors for p53-wildtype tumors) and PROTACs to degrade oncoproteins, while also targeting oncoprotein-stabilizing deubiquitinases (DUBs) like USP7/USP10 in advanced SCC. Selectivity must be addressed to spare normal tissue function. Advancing next-generation proteasome inhibitors that move beyond broad agents like Bortezomib to selectively target immunoproteasomes or exploit SCC-specific subunit dependencies. Designing rational combination therapies, including: Pairing DUB inhibitors with the targeted agents (e.g., EGFR/BRAF inhibitors) to overcome ubiquitination-mediated resistance. Combining ubiquitin pathway modulators with immunotherapy, such as inhibiting PD-L1-stabilizing DUBs to enhance checkpoint blockade. Creating stage-specific regimens, such as neoadjuvant/adjuvant therapies based on early signatures, and metastatic therapies targeting advanced evasion mechanisms. Leveraging the map for drug repurposing by screening existing compound libraries for agents that reverse prognostic DUP signatures or modulate SCC-relevant ubiquitin pathways.

4.5 Integration and systems medicine: Building the 3 PM framework

The ultimate realization of predictive, 3 PM for SCC necessitates *transcending isolated ubiquitinomics through multi-dimensional integration and dynamic systems medicine*. This requires *seamless multi-omic synthesis*, merging ubiquitinomics data with genomics (*FBXW7*/*TP53* mutations), transcriptomics (ubiquitin-altering splice variants), phosphoproteomics (PTM crosstalk), metabolomics (ubiquitin-driven Warburg shifts), and microbiome profiles (host-microbe ubiquitin interactions), leveraging *AI/ML* to decode complex datasets, identify master regulatory nodes, and build predictive models of tumor behavior. Complementing this, *computational modeling* must simulate the dynamic SCC ubiquitin-proteasome network across disease stages and therapeutic perturbations to forecast vulnerabilities and optimize intervention strategies. Crucially, *longitudinal ubiquitinome monitoring* via liquid biopsies enables real-time tracking of signature evolution during treatment and progression, facilitating adaptive therapy adjustments. Together, this integrated framework transforms static molecular snapshots into a living, actionable blueprint for precision oncology.

4.6 Addressing broader challenges and ethical considerations

To realize the transformative potential of ubiquitinomics in SCC management, future progress must proactively address four overarching challenges: *demographic diversity* requires actively recruiting ethnically, geographically, and lifestyle-diverse cohorts to map population-specific variations in the SCC ubiquitinome, ensuring globally relevant diagnostics/therapies. *Functional redundancy and resistance* in the UPS necessitates innovative strategies like multi-target inhibitors or adaptive therapy regimens to overcome tumors' rapid compensatory mechanisms. *Cost and accessibility* barriers must be dismantled through scalable technologies and equitable implementation frameworks to democratize advanced ubiquitination profiling and novel therapies; and *ethical frameworks* urgently need development to govern predictive data use, establishing guidelines for genetic predisposition disclosure, early-risk communication, and psychological support to navigate the implications of ubiquitinomics-driven prognostics. These pillars are non-negotiable for translating ubiquitinome discoveries into ethically sound, universally accessible 3 PM (predictive, preventive, personalized medicine).

4.7 Towards an ubiquitin-tailored future

The journey from the foundational ubiquitinome map of SCC to improved patient outcomes is complex but brimming with potential. The next decade will likely witness transformative advances in ubiquitin detection technology, leading to validated ubiquitination-based biomarkers for precise risk stratification (predictive medicine) and early detection (preventive medicine). Simultaneously, deep functional validation will unlock novel therapeutic targets within the UPS, paving the way for highly specific drugs and rational combination regimens (personalized medicine). Moreover, the vision is a future where an SCC patient's management is guided not just by histology and genomics but by a dynamic readout of their tumor's ubiquitination landscape. A biopsy or liquid sample could reveal the dominant ubiquitin-driven pathways, predicting aggressiveness and recurrence risk. Treatment could be tailored: neoadjuvant therapy targeting early-stage

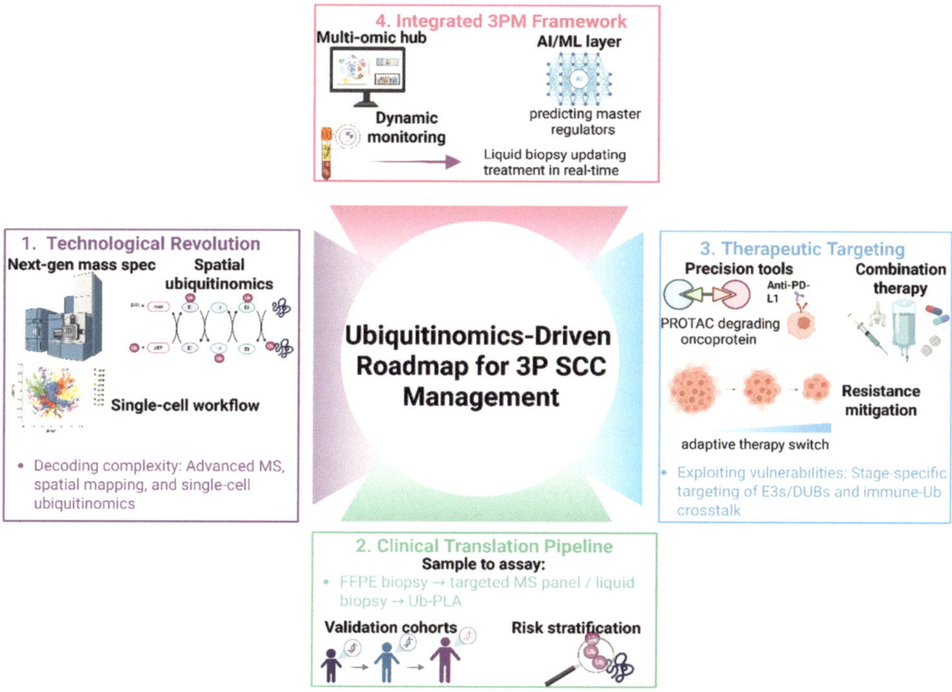

Figure 4.
Future prospects for SCC management: Technological advances, clinical translation, therapeutic targeting, and integrated 3 PM framework. This figure presents an Integrated 3PM (Precision, Predictive, Preventive Medicine) Framework for advancing cancer care. It is built upon a multi-optic hub and AI layer for data integration. The framework outlines a three-pillar strategy: 1) A Technological Revolution utilizing advanced mass spectrometry, spatial mapping, and single-cell analysis to decode tumor complexity. 2) A Clinical Translation Pipeline that processes patient samples (like biopsies) into actionable assays for validation and risk stratification. 3) Therapeutic Targeting with precision tools (e.g., PROTACs, immunotherapy) to exploit cancer vulnerabilities and dynamically adapt treatment, mitigating resistance.

ubiquitination drivers, surgery, followed by adjuvant regimens selected based on residual ubiquitination signatures, and metastatic therapy precisely inhibiting the ubiquitin-mediated resistance or immune evasion mechanisms identified in the individual's tumor. In addition, overcoming the substantial technical, biological, and clinical hurdles requires sustained interdisciplinary collaboration between ubiquitin biologists, oncologists, bioinformaticians, chemists, and clinical trialists. However, the reward, shifting SCC management from reactive to proactive, from generalized to exquisitely personalized, makes the pursuit of the ubiquitinomic horizon in SCC not just promising, but imperative for realizing the full potential of 3 PM. The work of Yang et al. has ignited this path; the future lies in traversing it with innovation and rigor (**Figure 4**).

5. Ethical and translational considerations for clinical implementation

The translation of ubiquitinomic discoveries from a research setting into routine clinical practice for SCC management presents significant ethical and practical challenges that must be proactively addressed to ensure equitable, effective, and responsible implementation within the 3 PM framework.

5.1 Ethical implications of predictive ubiquitinomics

The ability to stratify patients based on highly specific ubiquitination signatures raises several ethical considerations: Identifying patients with high-risk ubiquitinomic signatures, especially in pre-malignant or early-stage settings, could cause significant anxiety, "labeling," and psychological distress. Clear guidelines for communicating this complex prognostic information and providing integrated psychological support are essential. Consent processes must clearly explain the nature of ubiquitinomic testing, including its predictive uncertainty, the potential for discovering incidental biological findings, and how the data will be stored, used, and shared. Ubiquitinome data constitutes sensitive health information. Robust, transparent data governance frameworks are required to protect patient privacy and prevent misuse, such as discrimination by employers or insurers. There is a substantial risk that advanced ubiquitinomic profiling and subsequent targeted therapies will initially be available only at specialized, well-resourced centers, exacerbating existing health disparities. Deliberate strategies for cost-reduction and equitable resource allocation are necessary to prevent a "genomic divide" in SCC care.

5.2 Translational hurdles in biomarker and therapeutic development

Bridging the gap between ubiquitinomic discovery and clinical utility involves overcoming formidable translational barriers: Translating the 46 survival-associated DUPs into FDA-approved diagnostic tests is a major hurdle. This requires creating robust, cost-effective assays for FFPE biopsies (targeted MS panels, ubiquitination-specific immunohistochemistry) and liquid biopsies (detecting circulating ubiquitinated proteins or tumor-derived exosomes) and demonstrating that the assay reliably measures the ubiquitination signature (analytical validity) and that the signature accurately predicts the clinical outcome of interest (clinical validity) in large, prospective, multi-center SCC cohorts.

Proving that using the test to guide treatment decisions actually improves patient outcomes, a step that necessitates costly and lengthy randomized controlled trials. Developing drugs that target specific components of the ubiquitin system is fraught with challenges: E3 ligase inhibitors and DUB inhibitors must be highly specific to avoid disrupting the vast number of essential UPS functions in normal cells, which could lead to off-target toxicity (neurotoxicity). The UPS network is highly redundant and adaptive; tumors can rapidly develop compensatory mechanisms to bypass targeted inhibition, necessitating rational combination strategies from the outset. Clinical trials for UPS-targeted agents must incorporate ubiquitinomic biomarkers to identify the SCC patient subpopulations most likely to respond, moving beyond a one-size-fits-all approach.

5.3 Navigating the path to clinical integration

Successful integration of ubiquitinomics into SCC management requires a coordinated, multidisciplinary effort: Close collaboration between molecular biologists, clinical oncologists, pathologists, bioinformaticians, ethicists, and patients is crucial to design clinically relevant studies and implement findings. Regulatory bodies like the FDA will need to adapt to the evaluation of complex, multi-protein ubiquitination signatures as companion diagnostics, which differ from traditional single-analyte tests. Widespread adoption requires educating clinicians on interpreting ubiquitinomic reports and understanding the clinical implications of targeting the UPS, a system less familiar than canonical signaling pathways. While the ubiquitinome offers a revolutionary path for 3 PM in SCC, its journey to the clinic is inseparable from confronting these ethical and translational challenges. Addressing these issues head-on is not merely an administrative step but a fundamental prerequisite for realizing a future where SCC care is truly predictive, preventive, and personalized in an equitable and sustainable manner.

6. Conclusion

SCC emerges as a distinct clinical and molecular entity within colorectal cancer, characterized by unique ubiquitinomic profiles. The establishment of disease- and stage-specific ubiquitination patterns, with early stages defined by 385 DUPs facilitating oncogenic evasion and advanced stages by 223 DUPs driving metastasis and resistance, represents a transformative advance. Crucially, 46 survival-associated DUPs offer validated prognostic biomarkers and therapeutic targets, implicating key pathways like *Salmonella* infection, ferroptosis, and glycolysis.

However, the translation of this foundational knowledge into clinical 3 PM hinges on addressing two critical gaps: the functional validation of candidate DUPs to establish causal roles in pathogenesis, and large-scale, prospective validation studies to confirm the real-world prognostic and predictive power of ubiquitination signatures. Future progress depends on a concerted effort to advance ubiquitinomics technology, validate functional mechanisms, and develop UPS-targeted therapies. The definitive translational roadmap is clear: from ubiquitinome discovery, we must move through rigorous functional studies and prospective clinical validation to arrive at the final destination of clinically integrated, ubiquitin-informed predictive diagnostics, and personalized therapies for SCC.

Acknowledgements

This work was supported by the China National Nature Scientific Funds (82203592 to N.L.), Shandong Provincial Taishan Scholar Engineering Project Special Funds (NO. tstp20221143 to X.Z.), the Shandong Provincial Natural Science Foundation (ZR2022QH112 to N.L.; ZR2024QH606 to O.B.), and the China National High-end Foreign Expert Talent Program (H20240743; Foreign Expert: X.Z.). The authors also acknowledge the ChatGPT for language polishing.

Author's contributions

O.B. was responsible for design, formal analysis, and writing original draft. L.J was responsible for visualizations. H.S. provided the clinical samples, participated in the literature analysis, and reviewed the manuscript. X.Z. and N.L. were responsible for conceptualization, methodology, project administration, writing-review and editing, and critical revision. All authors proved the final manuscript.

Conflict of interest

We declare that the authors do not have competing interests.

Author details

Ousman Bajinka[1], Lamarana Jallow[2,3], Haiyu Shen[4], Na Li[1] and Xianquan Zhan[1,5]*

1 Shandong Provincial Key Laboratory of Precision Oncology, Shandong Cancer Hospital and Institute, Shandong First Medical University and Shandong Academy of Medical Sciences, Jinan, Shandong, China

2 Department of Biochemistry, Cell and Molecular Biology, College of Basic and Applied Sciences, University of Ghana, Legon, Accra, Ghana

3 Department of Biology, Division of Physical and Natural Sciences, School of Arts and Science, University of The Gambia, Banjul, The Gambia

4 Department of Colorectal Surgery, Shandong Cancer Hospital and Institute, Shandong First Medical University and Shandong Academy of Medical Sciences, Jinan, Shandong, China

5 Shandong Provincial Key Medical and Health Laboratory of Ovarian Cancer Multiomics, and Jinan Key Laboratory of Cancer Multiomics, Shandong First Medical University and Shandong Academy of Medical Sciences, Jinan, Shandong, China

*Address all correspondence to: yjzhan2011@gmail.com

References

[1] Derwinger K, Gustavsson B. Variations in demography and prognosis by colon cancer location. Anticancer Research. 2011;**31**(6):2347-2350

[2] Pham TT, Talukder AM, Walsh NJ, Lawson AG, Jones AJ, Bishop JL, et al. Clinical and epidemiological factors associated with suicide in colorectal cancer. Supportive Care in Cancer. 2019;**27**(2):617-621. DOI: 10.1007/s00520-018-4354-3

[3] Haji A, Ryan S, Bjarnason I, Donaldson N, Papagrigoriadis S. Colonoscopic high frequency mini-probe ultrasound is more accurate than conventional computed tomography in the local staging of colonic cancer. Colorectal Disease. 2012;**14**(8):953-959. DOI: 10.1111/j.1463-1318.2011.02871.x

[4] Loktionov A, Bandaletova T, Llewelyn AH, Dion C, Lywood HG, Lywood RC, et al. Colorectal cancer detection by measuring DNA from exfoliated colonocytes obtained by direct contact with rectal mucosa. International Journal of Oncology. 2009;**34**(2):301-311

[5] Yang H, Li N, Chen L, Zhou L, Zhou Y, Liu J, et al. Ubiquitinomics revealed disease- and stage-specific patterns relevant for the 3PM approach in human sigmoid colon cancers. The EPMA Journal. 2023;**14**(3):503-525. DOI: 10.1007/s13167-023-00328-2

[6] Mukund K, Syulyukina N, Ramamoorthy S, Subramaniam S. Right and left-sided colon cancers - Specificity of molecular mechanisms in tumorigenesis and progression. BMC Cancer. 2020;**20**(1):317. DOI: 10.1186/s12885-020-06784-7

[7] Wiencke JK, Zheng S, Lafuente A, Lafuente MJ, Grudzen C, Wrensch MR, et al. Aberrant methylation of p16INK4a in anatomic and gender-specific subtypes of sporadic colorectal cancer. Cancer Epidemiology, Biomarkers & Prevention. 1999;**8**(6):501-506

[8] Li ZW, Sun Q, Zheng Z, Yang LL, He L, Chen DN, et al. Clinicopathological features of colorectal amphicrine carcinoma. Zhonghua Bing Li Xue Za Zhi. 2022;**51**(8):708-712. DOI: 10.3760/cma.j.cn112151-20220106-00007

[9] Novikova MV, Rybko VA, Kochatkov AV, Khromova NV, Bogomazova SY, Dugina VB, et al. Izmenenie ékspressii membranoassotsiirovannykh belkov i tsitoplazmaticheskikh izoform aktina pri progressirovanii opukholeĭ tolstoĭ kishki cheloveka [A change in the expression of membrane-associated proteins and cytoplasmic actin isoforms in the progression of human colon tumors]. Arkhiv Patologii. 2017;**79**(2):15-21. DOI: 10.17116/patol201779215-21

[10] Liu LH, Lv H, Wang ZC, Rao SX, Zeng MS. Performance comparison between MRI and CT for local staging of sigmoid and descending colon cancer. European Journal of Radiology. 2019;**121**:108741. DOI: 10.1016/j.ejrad.2019.108741

[11] Chao S, Ying J, Liew G, Marshall W, Liew CC, Burakoff R. Blood RNA biomarker panel detects both left- and right-sided colorectal neoplasms: A case-control study. Journal of Experimental & Clinical Cancer Research. 2013;**32**(1):44. DOI: 10.1186/1756-9966-32-44

[12] Luo J, Zhou B, Zhao L, Yuan J, Zhou J, Shen L, et al. Circulating tumor DNA analysis detects micrometastatic disease and predicts recurrence

in a patient with colon cancer: A case report. Medicine (Baltimore). 2023;**102**(28):e34330. DOI: 10.1097/MD.0000000000034330

[13] Horiuchi A, Nakayama Y, Tanaka N, Kajiyama M, Fujii H, Yokoyama T, et al. Acute colorectal obstruction treated by means of transanal drainage tube: Effectiveness before surgery and stenting. The American Journal of Gastroenterology. 2005;**100**(12):2765-2770. DOI: 10.1111/j.1572-0241.2005.00276.x

[14] Sarela AI, Guthrie JA, Seymour MT, Ride E, Guillou PJ, O'Riordain DS. Non-operative management of the primary tumour in patients with incurable stage IV colorectal cancer. The British Journal of Surgery. 2001;**88**(10):1352-1356. DOI: 10.1046/j.0007-1323.2001.01915.x

[15] Chen TC, Liao YT, Huang J, Hung JS, Liang JT. Standardize the surgical technique and clarify the oncologic significance of robotic D3-D4 lymphadenectomy for upper rectum and sigmoid colon cancer with clinically more than N2 lymph node metastasis. International Journal of Surgery. 2024;**110**(4):2034-2043. DOI: 10.1097/JS9.0000000000001061

[16] Borchert DH, Schachtebeck M, Schoepe J, Federlein M, Bunse J, Gellert K, et al. Observational study on preservation of the superior rectal artery in sigmoid resection for diverticular disease. International Journal of Surgery. 2015;**21**:45-50. DOI: 10.1016/j.ijsu.2015.07.011

[17] Winter DC, Walsh R, Lee G, Kiely D, O'Riordain MG, O'Sullivan GC. Local involvement of the urinary bladder in primary colorectal cancer: Outcome with en bloc resection. Annals of Surgical Oncology. 2007;**14**(2):441-446. DOI: 10.1245/s10434-006-9144-3

[18] Weiss JM, Schumacher J, Allen GO, Neuman H, Lange EO, Loconte NK, et al. Adjuvant chemotherapy for stage II right-sided and left-sided colon cancer: Analysis of SEER-medicare data. Annals of Surgical Oncology. 2014;**21**(6):1781-1791. DOI: 10.1245/s10434-014-3631-8

[19] Fernández-Esparrach G, Bordas JM, Giráldez MD, Ginès A, Pellisé M, Sendino O, et al. Severe complications limit long-term clinical success of self-expanding metal stents in patients with obstructive colorectal cancer. The American Journal of Gastroenterology. 2010;**105**(5):1087-1093. DOI: 10.1038/ajg.2009.660

[20] Karim S, Brennan K, Nanji S, Berry SR, Booth CM. Association between prognosis and tumor laterality in early-stage colon cancer. JAMA Oncology. 2017;**3**(10):1386-1392. DOI: 10.1001/jamaoncol.2017.1016

[21] Zhang Y, Ma J, Zhang S, Deng G, Wu X, He J, et al. A prognostic analysis of 895 cases of stage III colon cancer in different colon subsites. International Journal of Colorectal Disease. 2015;**30**(9):1173-1183. DOI: 10.1007/s00384-015-2273-z

[22] Liao CK, Chiang JM, Tsai WS, You JF, Hsieh PS, Hung HY, et al. Primary tumor location in stage III colon cancer has prognostic impact on subsequent liver metastasis. Journal of Surgical Oncology. 2018;**118**(8):1301-1310. DOI: 10.1002/jso.25270

[23] Amri R, Bordeianou LG, Sylla P, Berger DL. Variations in metastasis site by primary location in colon cancer. Journal of Gastrointestinal Surgery. 2015;**19**(8):1522-1527. DOI: 10.1007/s11605-015-2837-9

[24] Huh JW, Kim YJ, Kim HR. Distribution of lymph node metastases

is an independent predictor of survival for sigmoid colon and rectal cancer. Annals of Surgery. 2012;**255**(1):70-78. DOI: 10.1097/SLA.0b013e31823785f6

[25] Ugai T, Akimoto N, Haruki K, Harrison TA, Cao Y, Qu C, et al. Prognostic role of detailed colorectal location and tumor molecular features: Analyses of 13,101 colorectal cancer patients including 2994 early-onset cases. Journal of Gastroenterology. 2023;**58**(3):229-245. DOI: 10.1007/s00535-023-01955-2

[26] Wang C, Wainberg ZA, Raldow A, Lee P. Differences in cancer-specific mortality of right- versus left-sided colon adenocarcinoma: A surveillance, epidemiology, and end results database analysis. JCO Clinical Cancer Informatics. 2017;**1**:1-9. DOI: 10.1200/CCI.17.00099

[27] Cortés-Guiral D, Pastor-Iodate C, Díaz Del Arco C, Del Puerto-Nevado L, Fernández-Aceñero MJ. CPEB4 immunohistochemical expression is associated to prognosis in stage IV colorectal carcinoma. Pathology, Research and Practice. 2017;**213**(6):639-642. DOI: 10.1016/j.prp.2017.04.020

[28] Guzmán-Gallego EA, Arias-Ortiz NE, Rodríguez-Betancourt JD. Colorectal cancer survival in Manizales, Colombia, 2008-2017: A population-based study. Revista Brasileira de Epidemiologia. 2023;**26**:e230040. DOI: 10.1590/1980-549720230040

[29] Hady HR, Sołdatow M, Lukaszewicz J, Luba M, Pierko J, Myśliwiec P, et al. Surgical treatment of malignant and benign colorectal neoplasms based on authors' clinical data. Advances in Clinical and Experimental Medicine. 2013;**22**(2):219-227